Engineering Materials

This series provides topical information on innovative, structural and functional materials and composites with applications in optical, electrical, mechanical, civil, aeronautical, medical, bio- and nano-engineering. The individual volumes are complete, comprehensive monographs covering the structure, properties, manufacturing process and applications of these materials. This multidisciplinary series is devoted to professionals, students and all those interested in the latest developments in the Materials Science field, that look for a carefully selected collection of high quality review articles on their respective field of expertise.

Indexed at Compendex (2021)

More information about this series at http://www.springer.com/series/4288

Annelise Kopp Alves
Editor

Technological Applications of Nanomaterials

Springer

Editor
Annelise Kopp Alves
Escola de Engenharia
Universidade Federal do Rio Grande do Sul
Porto Alegre, Rio Grande do Sul, Brazil

ISSN 1612-1317 ISSN 1868-1212 (electronic)
Engineering Materials
ISBN 978-3-030-86900-7 ISBN 978-3-030-86901-4 (eBook)
https://doi.org/10.1007/978-3-030-86901-4

This Springer imprint is published by the registered company Springer Nature Switzerland AG
The registered company address is: Gewerbestrasse 11, 6330 Cham, Switzerland

Preface

This book contains an overview of novel synthesis, characterization, and applications of nanomaterials. Based on an extensive state-of-the-art literature survey and the results obtained by our research group during the past years, this book presents techniques and special applications of classical and modern nanomaterials.

This book is aimed at students, researchers, and engineers who seek general scientific knowledge about nanomaterials with an application-oriented approach.

The following chapters present the general aspects of synthesis, characterization and innovative applications of different nanomaterials: erosion of CERMET nano-coatings (Chap. 1); thermoelectricity (Chap. 2); viral detection (Chap. 3); carbon quantum dots (Chap. 4); silica nanoparticles (Chap. 5); ballistic performance of armors (Chap. 6); ferroelectricity in nanoscale (Chap. 7); electrochromism (Chap. 8); Nb_2O_5 nanostructures (Chap. 9); magnetic hyperthermia (Chap. 10); inorganic pigments (Chap. 11); graphene (Chap. 12); microwave-assisted synthesis (Chap. 13); and applications of glycerol in the synthesis of nanomaterials (Chap. 14) and, nanomaterials and their influence in society through times (Chap. 15).

Porto Alegre, Brazil
July 2021

Annelise Kopp Alves

Contents

Jet Slurry Erosion of CERMET Nano-Coatings Obtained by HVOF

Freddy Galileo Santacruz Bastidas and Carlos Pérez Bergmann

Abstract Nowadays, the slurry erosion caused by solid particles has received considerable attention among researchers due to the intensity of the problems that it causes to equipment in service in the hydrometallurgical industry. Therefore, this has led to growing technological demand, focused on obtaining improvements in materials, especially in developing micro and nano-scale coatings, with higher resistance in service to more aggressive environments and subject to this type of erosion. Therefore, this chapter will address some of the materials and coatings studied that have excellent efficiency under extreme service conditions, such as slurry erosion and its synergy phenomena and some of the principal deposition methods nanometric powders obtention.

Keywords Nano-coating · Cermet · High energy milling · HVOF · Tribology · Wear · Slurry Erosion

Abbreviations

AISI	American Iron and Steel Institute
CERMET	Ceramic Carbide in Metallic Dies
PU	Polyurethane
PUR	Rigid Polyurethane Foam
HVOF	High Velocity Oxy-Fuel
PSP	Plasma Arc Spraying
PTA	Plasma Transferred Arc
O.M	Optical Microscopy
XRD	X-Ray Diffraction
SEM	Scanning Electron Microscopy
DIN	Deutsches Institut für Normung—International Organization for Standardization

F. G. S. Bastidas (✉) · C. P. Bergmann
Universidade Federal do Rio Grande do Sul, Porto Alegre, Brazil
e-mail: galileo.santacruz@ufrgs.br

A. Kopp Alves (ed.), *Technological Applications of Nanomaterials*,
Engineering Materials, https://doi.org/10.1007/978-3-030-86901-4_1

A.S.T.M.	American Society for Testing and Materials
D	Crystallite size
HV	Vickers hardness
MA	Mechanical Allowing
MM	Mechanical Milling
rpm	Revolutions per minute

1 Introduction

In the hydrometallurgical industry and their applications, such as hydraulic turbines in hydroelectric plants, in the petrochemical industry, in civil construction, in the processing of minerals, slurry pumps, pipes, among others, the loss of metal due to erosion by slurry caused by the particulate material in the liquid represents a major industrial problem, affecting the life of the components and reducing their performance. Thus, the ideal choice of engineering materials is crucial to decrease wear and improve their tribological behavior [1, 2].

In the study for improvements in these conditions, the technology of surface treatments and nano-coatings is important because it understands and studies the mechanisms generated on the surface because most of the failures that occur due to erosion and corrosion depend on them [3, 4]. Thus, to optimize its properties and, consequently, its performance in operation, mechanical and surface studies of the different types of thin-film coatings are essential [5–9].

Among the different alternatives for such applications that are currently being studied there are austenitic and martensitic stainless steels [10–12]. Austenitic stainless steels are used in many components where corrosion resistance is crucial. However, under the mechanical action of hard particles, they present a high plastic deformation and wear [13].

On the other hand, martensitic stainless steel has better mechanical resistance to erosive particles than austenitic steel, but its corrosion resistance is lower [14, 15].

The cermet (ceramic carbides in metal matrices) combine different materials in micro and nanostructural scale results in materials with peculiar properties and improved performance, which are not shown by the individual constituents. Based on this conception of composite materials, the cermet, which by the set of properties, has shown excellent results in resistance to erosive and abrasive wear [16].

Cermet nano-coatings widely used, such as WC–Co, Cr3C2–NiCr, WC–(W, Cr)2C–Ni, WC–NiCrBSi, belongs to one of the most important groups of materials obtained by thermal spray techniques into coatings and deposited from nanostructured powders, which are predominantly applied for the protection against wear, such as abrasion, sliding, and slurry erosion [17]. Improvements in the performance of these coatings are frequently achieved with the use of nanostructured materials. In addition, the physical and mechanical properties of these materials are superior due to the reduced size of grain and the high volumetric fraction of atoms present at the grain boundaries [18].

To apply these coatings over substracts, the different techniques in surface engineering are valuable; many scientists usually employed processes thermal spraying techniques use flame combustion, high-velocity oxy-fuel (HVOF), detonation, electric arc systems, propelled electric arc, plasma systems with transferred arc (PSP-Plasma Arc Spraying), and non-transferred (PTA- Plasma Transferred Arc). The HVOF technique has been shown to obtain very hard coatings with excellent adhesion and cohesion properties. Coatings applied by the arc-spray technique have also been used; however, this technique resulted in a coating characterized by large-sized lamellar structure and high porosity, limiting their performance [19, 20].

WC–Co-based cermets provide abrasive wear resistance properties. In addition, the corrosion resistance of these nano-coatings can be increased with the addition of a small amount of Cr. The composition WC-10% Co-4% Cr is a set widely considered nano-coating in hydrometallurgical industries applications and their components to extend their useful life [16]. WCCoCr nano-coatings are commonly deposited through the high-velocity thermal spray (HVOF—High-Velocity Oxy-Fuel) process [21, 22].

2 Martensitic Stainless Steel and Industrial Applications

Among the coated materials, some of the most relevant ones are martensitic stainless steels, given their wide industrial applications. Stainless steels are a particular class of steel alloys known mainly for their corrosion-resistant properties. The characteristics of stainless steel associated with these alloys are obtained through the formation of a film on the surface of invisible and adherent chromium-rich oxide that, when damaged, scratched, or eroded by some element or type of machining, has the rare ability to repair it if in the presence of oxygen.

Chromium is the alloying element that gives corrosion resistance to stainless steel. Still, many other elements can stabilize different phases, provide corrosion resistance and produce improved mechanical properties. Mainly they are alloys of iron, carbon, chromium, nickel, molybdenum, nitrogen, titanium. In general, according to the chemical composition of the alloy, there are four main types of stainless steel: ferritic, martensitic, austenitic, and duplex [23]. These alloys have been a source of study since the first world war, in that they provide a variety of mechanical properties for many applications [24].

Martensitic stainless steels are the first branch of chromium steels and can be heat treated by quenching and tempering. They were the first to develop industrially and represent part of the 400 AISI series (American Iron and Steel Institute) with very important properties such as heat treatment hardening. Therefore, they can develop high mechanical strength and hardness levels, have easy machining, good corrosion resistance, and have magnetic properties [25, 26].

Martensitic stainless steels are high-carbon-content alloys (0.1–1.2%) with chromium (12–18%) due to its tribological properties in increasing the useful life of its components with high mechanical resistance such as tensile, creep, fatigue, high yield stress, better weldability, and moderate corrosion resistance, both at high and low temperatures (up to approximately 873 K), are an excellent option for industrial applications [14, 27]. For this reason, these steels are typically used in hydroelectric plants, automotive industry, distillation towers, exploration platforms in the petrochemical industries, manufacture of nuclear energy equipment and reactors, steam, and gas turbines, cutting tools, civil construction, compressors, components in the chemical industry, mineral processing and better mechanical resistance to erosive particles than austenitic steel, among others [4, 10, 11]. Alloys hardened by heat treatment also contain properties that allow its use in applications such as containers for storage of organic acids or saline solutions [15].

The phase transformations in martensitic stainless steels with low carbon contents have been studied extensively for decades because, as a result of the heat treatments carried out on the material, a dispersed percentage of austenite is obtained in its microstructure, which depending on the present, provides the material with specific mechanical properties for the manufacture of components subject to high demands. Therefore, controlling the amount of austenite obtained after heat treatments in the steel is essential for the piece's performance [28].

The manufacture of alloy steels resistant to erosion and for tools is mainly based on steels bonded with strong elements that form carbides, such as Cr, Mo, W, and V. These elements are divided between the carbides and the austenitic matrix during solidification, hot work, annealing and austenitizing before hardening. During hardening, the bonded carbides formed in austenite are retained, and the austenitic matrix is transformed into martensite. Subsequently, the alloying elements partition during the tempering as the transformations of retained austenite and the fine alloy carbides precipitate in the as-quenched martensite. Thus, all microstructure elements provide hardening and wear resistance: retained carbides, tempered martensite, and carbides formed during tempering. The final microstructure is a martensitic matrix with small fractions of delta ferrite and stable austenite [29, 30].

The resistance to wear in this type of steel depends considerably on the increase in the volumetric fraction of carbides as well as on the increase in their hardness [14].

A typical outcome of quenching and tempering thermal treatments is shown in Fig. 1 [25]. The martensitic stainless steel (AISI 410) samples were austenitized at 1263 K for 35 min, then oil-quenched and tempered at 793 K for 35 min afterward. The microstructure obtained after this procedure was composed of martensite with some precipitated carbides on the grain contours.

All martensitic stainless steels can be tempered and quenched, and the hardness achieved will depend on the carbon content of the alloy. In low carbon steels, which contain about 12% Cr and 0.1% C, it can have a variety of hardness levels, from 20 to 40 HRC, and in high carbon steels, the hardness can reaching values close to 60 HRC. As with carbon steels, these alloys are susceptible to fragility when subjected to heat treatment after hardening in the range of 723 to 813 K.

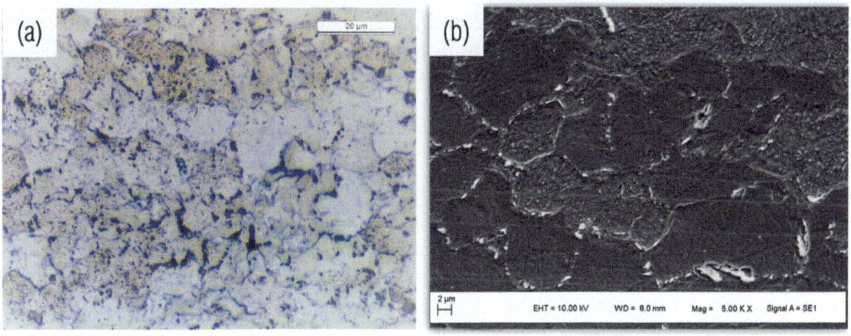

Fig. 1 Typical resulting microstructure of AISI 410 stainless steel after thermal treatments by **a** LM 1000×, and **b** SEM 5000×, after Vilella acid etching. Reproduced with permission from [25]

Santacruz et al. [25] show the results of Vickers microhardness indentation measurements on the cross-section (HV0.3–2.94 N), performed on the thermally treated martensitic stainless steel, obtaining values of 219 HV0.3, which is very similar to that reported by Maiti et al. [31] of 199 HV0.3.

3 Tungsten Carbide (86WC-10Co-4Cr) Cermet Nano-Coatings

3.1 Cemented Carbides (CERMET)

Cemented carbides, alloys also known as hard metal, are powder metallurgical materials obtained from finely fragmented powders of hard particles of refractory carbides, usually tungsten, but also titanium, tantalum, niobium, chromium, molybdenum, and vanadium. These are sintered with one or more metals in the iron group (iron, cobalt, chromium, or nickel), which form the binding phase and form an alloy of high hardness, high resistance to compression, wear, and resilience. The properties of carbide can be modified by varying the percentage of the binder phase and hard particles and the size of these [32, 33].

The cermet composite (ceramic–metal) is a hard metallic material that combines the high hardness, wear-resistance, and chemical stability of the ceramic phase (WC tungsten carbide), with ductility and toughness of the metallic phase (Co cobalt), Co is Widely used as a metallic binder, due to its high wettability and good solubility with WC, and good mechanical properties [34].

WC–Co cermets are formed by a homogeneous distribution of faceted WC grains incorporated in a Co matrix, whose properties depend mainly on the composition, microstructure and chemical purity of carbides. The excellent wear resistance of these materials, which exceeds between 12 and 20 times the strength of high rapid steel, is due to its unique combination of high hardness and moderate levels of fracture toughness [35].

3.2 *Nanostructured Powder*

Many studies about WCCo coatings have been using conventional commercial spray powder material, but currently, the literature presents more and more investigations on the use and performance of spray powders in coatings that use nanostructured WCCoCr in their composition [36–39].

A great improvement in performance in terms of physical and mechanical properties with better wear resistance, mainly to the slurry erosion in coatings (WCCo) obtained by thermal spray techniques, can be achieved using nanostructured materials. Furthermore, these nanocomposite materials' physical and mechanical properties are superior because of the reduced grain size and the high volumetric fraction of atoms present in the grain boundaries [40]. Samely, these nanocomposites have been shown high strength, fracture toughness, stiffness, and high-temperature properties are potential candidate materials for structural applications in aerospace hardware, turbo engine parts, turbines, and pumps.

The possibility of using the high-energy ball milling technique to obtain nanopowders, controlling their characteristics through process parameters, is currently of great interest for researchers in the development of coatings applied by thermal spray, with a protective function against wear and tear and erosion.

MA is a high-energy ball milling technique used to produce a variety of materials, such as borides, carbides, nitrides, oxides, and nanocomposites [41–44]. Many studies have been conducted about WCCo nanocomposites obtained by high-energy ball milling [45–51].

3.2.1 Mechanical Alloying (MA)—High-Energy Ball

High energy ball milling (MA) is a solid-state material synthesis technique that involves repeated deformation cycles, cold welding, and fracturing of powder particles in high-energy mills [52]. The method has important advantages, for example, it is very easy to obtain nanostructures, a large volume fraction of the reinforcement phase can be introduced into the composite, and consolidation of the ground powder is relatively easy.

The process was developed by John Benjamin and his colleagues at the Paul D. America Research Laboratory of the International Nickel Company (INCO). The technique resulted from long research into the production of a high-temperature resistant nickel superalloy for application in gas turbines. The required corrosion and oxidation resistance has also been included in the alloy with the appropriate addition of alloying elements. The technique developed by Benjamin was called milling/mixing, but INCO patented it under the name Mechanical Alloying (MA), and this is the term commonly used in the literature [52].

Mechanical alloying is the generic term used to describe the processing of metal powders in high-energy mills. However, depending on the initial state of the powder mixture and the processing steps involved, different terms have been used in the literature. Two terms are most commonly used: Mechanical Alloying (MA) and Mechanical Milling (MM). The process is called MA when mixtures of powders (of different metals or alloys/composites) are milled together. Therefore, if pure metal powders A and B are milled together to synthesize an alloy, the process is designated as MA. The material transfer is involved in this process to obtain a homogeneous alloy. On the other hand, when powders with uniform composition (often stoichiometric), such as pure metals, intermetallics, or alloys, are processed in a high-energy mill, and there is no material transfer in the homogenization, the process is called Mechanical Milling (MM). The advantage of MM over MA is the short processing time since the powders are bound, and only a reduction in particle size and/or other necessary transformations are mechanically induced. An additional advantage of MM concerning MA is that MM reduces the oxidation of the powders' constituents due to the reduced processing time [52].

Due to its attributes, Mechanical Alloying has been employed to produce a variety of commercially useful and scientifically interesting materials. We can highlight attributes such as the production of finely dispersed second-phase particles (usual oxides), the extension of the solid solubility limit, the refinement of the grain size to the nanometric scale, the synthesis of new crystalline and quasi-crystalline phases, the development of amorphous (vitreous) phases, the disordering of ordered intermetallics, the possibility of forming alloys of elements with difficult miscibility as well as the induction of chemical reactions at low temperatures [42].

3.2.2 Mechanism of Alloying (MA)

The powder particles are repeatedly flattened during the high-energy milling process, cold-welded, fractured and cold-welded again. A collision between two grinding bodies traps a small amount of dust between them Fig. 2. Typically, circa1000 particles weighing 0.2 mg are trapped during each collision. The impact force plastically deforms the powder particles, causing hardening and fracture. The new surfaces created allow particles to weld together, leading to an increase in particle size. As the particles have low hardness in the early stages of grinding, the tendency for them to weld together and form large particles is high. Particles at this stage have a characteristic lamellar structure, consisting of various combinations of the initial constituents. As deformation continues, the particles harden and fracture by a fatigue mechanism. Fragments generated by this mechanism can continue to reduce their size in the absence of strong agglomeration forces. At this stage, the tendency to fracture predominates over cold welding. Due to the continued impact of the grinding

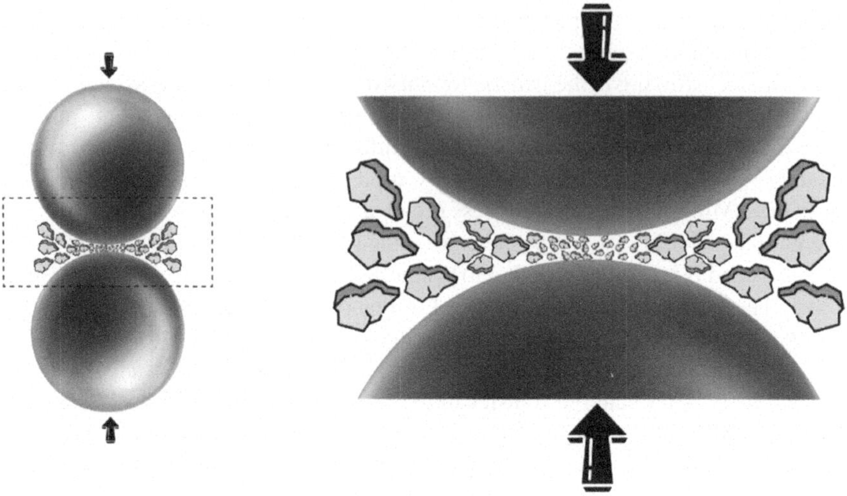

Fig. 2 Ball–powder–ball collision of powder mixture during mechanical alloying

bodies, the structure of the particles is constantly refined, but the size of the particles remains the same. Consequently, the interlamellar distances decrease, and the number of lamellar structures increases in each particle [53].

After a certain milling time, a steady state of equilibrium is reached when there is a balance between the welding rate, which tends to increase the average particle size, and the fracture rate, which tends to decrease the average particle size of the composite. Smaller particles can withstand deformation without fracturing and tend to weld into larger parts, which tends to form particles of intermediate size. The particle size distribution at this stage is narrow because while large particles tend to reduce their size, small fragments tend to grow by agglomeration Fig. 3 [52].

The specific time required to develop a given structure in a system will function the initial particle size and ingredient characteristics, equipment used, and equipment operating parameters. However, in many cases, the rate of refinement of the internal structure (particle size, crystallite size, interlamellar spacing, etc.) is an approximately logarithmic function with processing time, with the initial particle size being relatively important. Figure 4 [52] shows that between a few minutes and an hour, the interlamellar spacing becomes small, and the crystallite (or grain) size is refined to nanometric dimensions. The ease with which nanostructured materials can be synthesized is one of the reasons MA is extensively employed to produce nanocrystalline materials.

3.2.3 Raw Materials for Mechanical Alloying (MA)

The raw materials used for MA are commercially available pure powders that have particle sizes in the range of 1 ± 200 mm. Materials are typically pure metals,

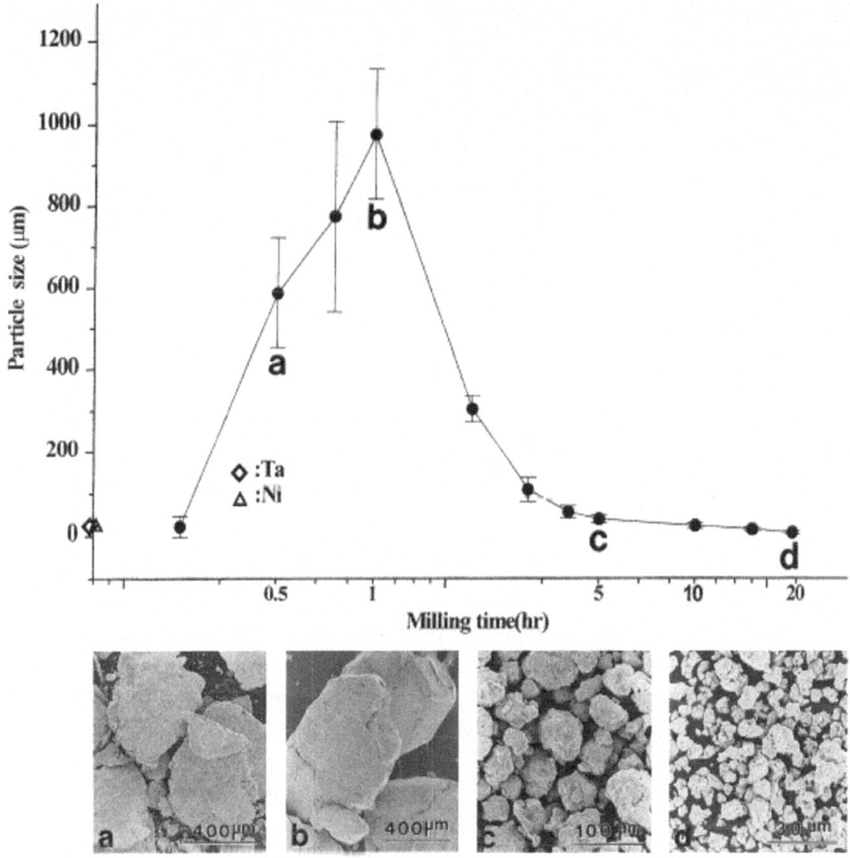

Fig. 3 Narrow particle size distribution caused by the tendency of small particles to weld together and large particles to fracture under steady-state conditions. Edited and reproduced with permission from [52]

metal alloys, pre-alloyed powders, carbides, nitrides, and oxides. Initially, the powder charge consisted of at least 15% by volume of a ductile phase to act as a binder. However, in recent years, mixtures of brittle materials have been processed, resulting in the formation of alloys [54]. Thus, it is possible to process mixtures of ductile-ductile, ductile–brittle, and brittle-brittle powders to produce new alloys through high-energy milling.

3.2.4 Influence of Mechanical Alloying (MA) Parameters in the Nanopowders Obtention Process

The transfer of energy from the milling bodies to the powders to be synthesized is affected by several parameters, which directly influence the final characteristics of

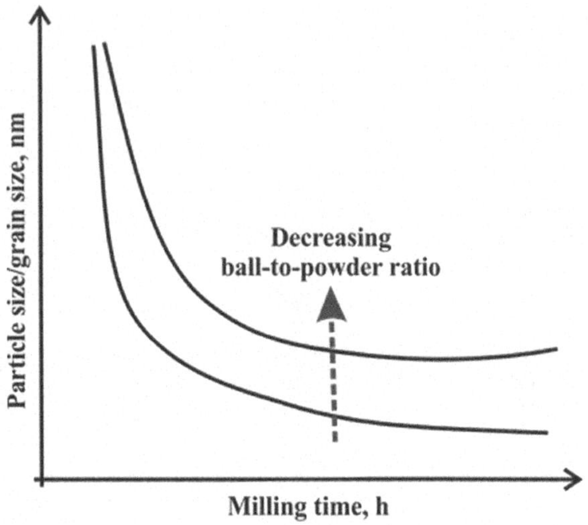

Fig. 4 Refinement of particle and grain sizes with milling time. Edited and reproduced with permission from [52]

the processed powders. According to Suryanarayana et al. [52], the most important parameters to take into account according to the purpose and its application are the types of mill, milling container, milling energy/speed, milling time, type, size, and size distribution of grinding medium, ball-to-powder weight ratio, the extent of vial filling, milling atmosphere Process control agent, the temperature of milling.

3.2.5 Types of Mills Available for Mechanical Alloying (MA) in the Nanopowders Obtention Process

Different types of high-energy milling equipment are used to produce powders through MA. They differ in capacity, milling efficiency, impact energy of the grinding bodies, and the possibility of controlling the milling temperature. Vibrating mill, attritor, and planetary mill are the most popular for carrying out MA experiments.

Vibratory mills are widely used in laboratories for comminution and synthesis of materials by MA. They can grind a small amount of material (among 10 and 20 g) at a time. The cycle combines lateral movements and forward and backward movements, describing the shape of a figure eight. At each cycle, the balls collide with each other and with the container's walls, which entails grinding and mixing the material. Because of the amplitude (approximately 50 mm) and speed (1200 rpm), the speed of the grinding bodies is high (on the order of 5 m/s), and the impact force is high. Therefore, there are various materials for this type of mill container, including stainless steel, tungsten carbide, alumina, zirconia, agate, and some polymers [52].

The planetary mill gets its name due to the rotation and translation movement of the grinding vessel, similar to the movement of the planets. The support disc moves, and the jug rotates on its axis. The centrifugal force produced by the combined rotation/translation movement causes the grinding bodies to interact with each other and the container's walls. The impact of the grinding bodies against each other considerably intensifies the impact importantly. Although the linear speed of grinding bodies in this type of mill is as high as in vibrating mills, the impact frequency is much lower. Therefore, planetary mills are considered to be low-energy grinding mills compared to vibrating mills.

The attritor mill is composed of a vertical cylinder with a central axis that has horizontal rods distributed along the length of the central axis. This mill can grind large amounts of material at a time (from 0.5 to 40 kg). However, the speed of the milling medium is much lower (circa 0.5 m/s) than in planetary and vibrating mills; consequently, the grinding energy is low. Milling containers are made of stainless steel or stainless steel coated on the inside with alumina, silicon carbide, zirconia, rubber, and polyurethane. A variety of grinding media are also available—glass, mullite, silicon carbide, alumina, zirconia silicate, zirconia, stainless steel, and tungsten carbide.

3.2.6 Types of Milling Container Available for Mechanical Alloying (MA) in the Nanopowders Obtention Process

The material in the container can break off and become incorporated in the powder due to the impact of grinding bodies on the walls of the container. If the container material is different from the powder, the powder may be contaminated with the container material. On the other hand, if the materials are the same, the chemical composition of the powder can be changed [52].

The materials commonly used in grinding jars are stainless steel, hardened steel, tool steel, and WCCo. Materials such as alumina, agate, yttria-stabilized with zirconia, titanium, and porcelain are used in specific cases.

3.2.7 Milling Energy/Speed for Mechanical Alloying (MA) in the Nanopowders Obtention Process

The grinding speed influences the energy of the grinding bodies. The faster the mill spins, the greater the energy transferred to the material to be milled. However, there are limitations on the maximum speed that can be employed. At speeds above the critical speed, the grinding media adheres to the container walls and does not exert impact force on the material. Thus, the maximum speed must be very close to the critical speed for the grinding bodies to fall as high as possible and produce the maximum impact energy.

3.2.8 Milling Time for Mechanical Alloying (MA) in the Nanopowders Obtention Process

Milling time is one of the most important process variables. The time is determined according to the type of mill used, the grinding intensity, the BPR, the grinding temperature, and the system to be ground. It should be noted that the level of contamination increases, and the formation of undesirable phases can occur for long grinding times.

3.2.9 Grinding Medium for Mechanical Alloying (MA) in the Nanopowders Obtention Process

The density of the grinding bodies must be high for the balls to create sufficient impact force on the powder. The most used materials are stainless steel, tempered steel, and WCCo. Wherever possible, it is desirable that the grinding jar and grinding bodies are of the same material to avoid cross-contamination of the ground material.

The size of the grinding media also influences the grinding efficiency. The larger the size of the balls, the greater the transfer of impact energy to the powder particles. Most researchers use single-size grinding media. However, studies have shown that higher collision energies are obtained with different sizes [55, 56].

Grinding Medium for Mechanical Alloying (MA) in the Nanopowders Obtention Process

Ball-to-powder weight ratio—BPR significantly affects the time needed to obtain a certain phase in the ground material. The higher the BPR, the shorter the time required. As the proportion of balls increases, the number of collisions per unit of time increases, and consequently, more energy is transferred to the powder particles, making the alloying process faster.

The Extent of Filling the Vial for Mechanical Alloying (MA) in the Nanopowders Obtention Process

Since the grinding of particles occurs due to the impact force exerted on them, there needs to be enough space for the balls and powder to move freely in the grinding vessel. Therefore, if the amount of balls and powder is too small, the production rate will be too low. On the other hand, if the quantity is large, there will not be enough space, and the collision energy will be less. Usually, less than 50% of the container capacity is used.

Milling Atmosphere for Mechanical Alloying (MA) in the Nanopowders Obtention Process

The most significant effect of the grinding atmosphere is powder contamination. Therefore, the materials are ground in jars under a vacuum or filled with inert gases such as argon or helium. High purity argon is the most used atmosphere to prevent oxidation and/or contamination of powder. Different atmospheres have been used during grinding for specific purposes. For example, nitrogen or ammonia has been used to produce nitrides. The presence of air in the grinding vessel leads to the production of oxides and nitrides in the powders, especially if the material is reactive in nature.

Process Control Agent (PCA) for Mechanical Alloying (MA) in the Nanopowders Obtention Process

Process control agents (also called lubricants or surfactants) are added to powders during milling to reduce the effects of cold welding. PCAs can be solids, liquids, or gases. These are commonly organic compounds that adsorb on the surface of particles and minimize cold welding. The most important PCAs are stearic acid, hexane, methanol, and ethanol. The choice of grinding process control agent depends on the nature of the powder to be ground and the purity of the final product desired. The nature and amount of PCA used during milling determine the final particle size and powder yield. The use of larger amounts of PCA usually reduces the particle size by 2 to 3 orders of magnitude [52].

The Temperature of Milling for Mechanical Alloying (MA) in the Nanopowders Obtention Process

The grinding temperature is another important parameter that influences the constitution of the ground material. Since diffusion processes are involved in forming crystalline phases, amorphous phases, solid solutions, intermetallics, and nanostructured materials, the temperature is expected to significantly affect these processes. If the generated temperature is high, the associated high diffusivity will lead to processes that will result in recovery and recrystallization. In this case, a stable phase will form, such as an intermetallic. On the other hand, if the temperature is low, recovery will be lower, and an amorphous (or nanocrystalline) phase will be formed.

3.3 Thermal Spray Coatings

The coatings applied by thermal spray or spray technology are formed through the deposition with high velocity and impact energy of successive layers of splats from the melting of the starting material in the form of powder, wire, or rod (millions of particles per cm^2/s), which flatten and solidify on a cold substrate surface, resulting in a thin film macro structure known as lamellae or splats that is mechanically anchored to surface irregularities [57]. The typical structure of thermal spray coatings is a cohesive adhesion of sheets of the sprayed material in combination with inclusions

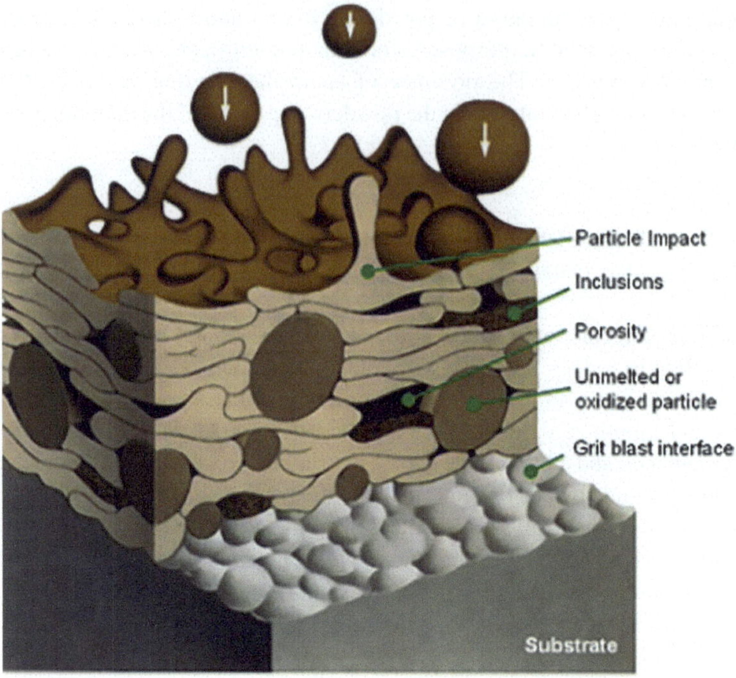

Fig. 5 Schematic cross-section of a typical microstructure of thermally sprayed coatings, showing the lamellar coating build-up with pores, oxides, and unmelted particles. Reproduced with permission from [60]

of oxide, micro-cracks, solid particles, and porosity [58, 59]. Figure 5 shows the cross-section of these types of coatings [60].

The physical characteristics and behavior of the applied layers depend on the cohesive forces between the films, the size, and the morphology of the porosity, cracks, and defects in the ultra-fine and granular microstructure within them [61]. A lamella or splat is the result of a drop of molten material, approximately ten to one hundred micrometers in diameter, is melted by the flame and impacts on the surface, where it cools quickly, flattens, and solidifies, with a cooling rate of approximately between 106 and 108 m/s [62].

The aspects of the formation and solidification of the layers are complex. They are also linked to the shape of the thin films, which is strongly influenced by the spraying angle because it has an important effect on the characteristics of the coating, such as porosity, microhardness, and the same efficiency [63].

The variation in the application parameters of the thermal spray coating influences its properties, in terms of the size and distribution of the porosity, oxide content, residual stresses, cracks in macroscopic and microscopic scales, which have a direct impact on the life and quality of the products themselves [64, 65].

To obtain good adhesion of the material projected on the base metal, a previous preparation must be carried out, eliminating all contaminants and maintaining the surface roughness. The main mechanism of adhesion of the coating to the substrate is mechanical anchoring [66]. In addition to this preparation, it depends on the thickness of the coating to be obtained, the nature of the projected material, and the piece's shape to be coated. Therefore, proper preparation of the substrate before applying the coating is very important because it significantly influences the bond strength and adhesion of the coating to the substrate [67].

The adhesion between the thermal spray coating and the substrate can be mechanical, metallurgical, chemical, physical, or a combination of these forms [68]. In addition, adhesion is influenced by some factors, such as the coating material, the condition of the substrate, the surface roughness, cleanliness, the surface temperature before, during and after thermal spraying and the impact velocity of the particle [69, 70].

Many metals, ceramics, inter-metallic composites, some polymers, and glasses can be deposited by one or more of the thermal spray processes on different types of substrates, such as metals, oxides, ceramics, glasses, and polymers.

The proper selection of the thermal spray process is determined by the desired material such as coating, performance requirements, part size, shape, and economic factors. Thermal spray techniques utilize flame combustion systems (HVOF), detonation, electric arc systems, propulsion electric arc, plasma transferred arc (PSP), and non-transferred (PTA) systems. The first system uses flue gases as a heat source. The second and third systems are constituted by electrical energy processes as heat sources [60].

Coating materials can be applied using several different processes. The thermal coating methods use fuel combustion, plasma spray, and electric arc systems. The coatings can be applied in standard atmospheric conditions or specialized and highly controlled atmospheric environments, even underwater. Coatings can be applied manually or with the automated precision of software-driven robotics. The thermal spray process is mainly classified by feedstock material, thermal energy source and the kinetic energy produced. Figure 6 shows the flowchart of the types of thermal spray processes.

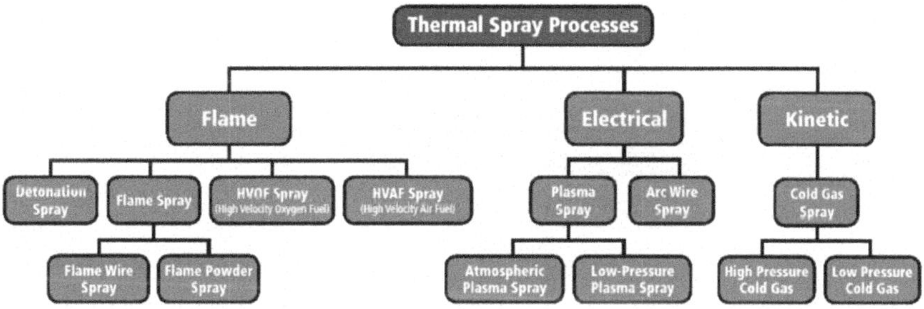

Fig. 6 Types of thermal spray processes

3.4 HVOF Thermal Spray

One of the most used industrial processes to mitigate the harmful effects of slurry erosion is HVOF (High-Velocity Oxy-Fuel). HVOF was first developed by Union Carbide (Thorpe Richter 1992), nowadays known as Praxair Tech. HVOF was developed after great efforts to produce thermal spray coatings with better mechanical properties than those obtained by conventional flame spray processes. The HVOF process was commercially introduced in the 1980s by JetKote (Deloro Stellite, Inc.) [71] and, after more than 40 years in the market, is currently one of the most preferred thermal spray processes for industrial applications [5]. Applied in WC–Co coatings, the wear rate can be reduced up to 1000 times compared to some stainless steel depending on the impact angle [72].

HVOF thermal spray is a process designed to accelerate molten particles to extremely high speeds, as their combustion of a mixture of fuel (kerosene, propane, propylene, hydrogen, or acetylene) and oxygen takes place in a small chamber, resulting in a discharge of carbon dioxide gases, water vapor and thermal expansion due to the depletion of these gases.

Due to the high pressure created in the combustion chamber, the gases generate supersonic velocities (gas velocities of 2000 m/s were observed), accelerating the molten particles. Although the particles do not reach the velocity generated by the gases, they do reach very high velocities of approximately 750 m/s. Figure 7 schematically illustrates the HVOF thermal spray process, and Fig. 8 shows the equipment in operation.

Figure 9 shows specimens of the thermally sprayed 86WC-10Co4Cr (Woka 3653) coating with fine spherical powders from Oerlikon Metco (Pfaffikon, Switzerland) with a particle size distribution of nominal range 45 + 11 μm and apparent density 4.8 5.8 g/cm^3 [75] in the form of a cylinder with 30 mm in diameter and 10 mm in height. Then, they were cut using a precision waterjet cutter. No additional finishing is required. The process does not cause distortions in the material of mechanical and thermal origin (since there is no heating), which ensures a clean-cut, and in the case of sheets coated by thermal spray, without detachment from it. The approximate chemical composition of the WOKA-3653 powder is given in Table 1. HVOF spraying parameters are briefly listed in Table 2.

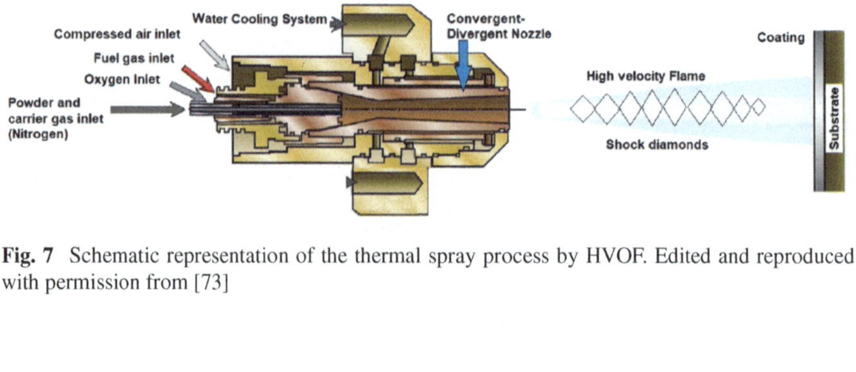

Fig. 7 Schematic representation of the thermal spray process by HVOF. Edited and reproduced with permission from [73]

Fig. 8 Thermal spray equipment used for coatings in operation. Adapted and reproduced with permission from [74]

Fig. 9 Samples of the 86WC-10Co4Cr coating (Woka 3653) thermally sprayed in the form of a cylinder with 30 mm in diameter and 10 mm in height. Reproduced with permission from [74]

Table 1 Chemical composition of WOKA-3653 (wt.%) [75]

Elements	Compositon (%)
W	Balance
Co	8.5–11.5
Cr	3.4–4.6
C (total)	4.8–5.6
Fe (max)	0.2

Table 2 Coating parameters of HVOF equipment [75]

Spray parameters	Value
Propylene flow rate	4620 L/h
Oxygen flow rate	15,180 L/h
Powder feed rate	42 g/min
Spray distance	230 mm

One of the most important advantages of the HVOF technique concerning other thermal spray processes, including the plasma type, is the high velocity achieved by the particles and the low temperatures involved (coatings with lower carbide decomposition than other deposition processes), which minimizes damage to the coating and substrate [76].

Nano-coatings produced by HVOF, especially with WC base, is hard, with low porosity, a homogeneous microstructure and reduced chemical decomposition, dense and of high quality, exhibiting a high adhesion (some above 12,000 PSI or 83 MPa) to the substrate, low oxidation and excellent wear resistance for greater durability and profitability of the components [77].

Some parameters are very important for the properties of nano-coatings deposited by HVOF, such as substrate preparation and temperature; composition, morphology, and size distribution of nanopowders; control of gas flow, relative movement of the gun and substrate, deposition angle, among others [22].

All of this contributes to WC-CoCr cermet nano-coatings sprayed generally with HVOF being able to survive the harshest service conditions, particularly in wear applications in slurry and corrosive media, greatly increasing service life. Figure 10 [26] shows the micrograph of the 86WC-10Co4Cr coating obtained by HVOF thermal spray. This one has a low amount of oxides, with a continuous and microstructurally homogeneous shape, with a low presence of cracks and pores. However, it is not possible to distinguish lamellar structure. This can be explained by the fusion of WC-Co-Cr particles, which induces a better distribution in the sprayed material [78].

The average thickness of the HVOF 86WC-10Co-4Cr coating is circa 227 μm. Santacruz et al. [26] calculated using micrographs of cross-sections processed using ImageJ software, and 100 to 200 measurements were performed on different samples to obtain their arithmetic mean. Likewise, the researchers obtained an average porosity of 0.7%, which is considered acceptable within the values presented in the literature and according to the criterion specified by the manufacturer Oerlikon Metco of a mean porosity lower than 1.0%. The procedure was realized through digital image analysis according to ASTM E2109-01 [79] in three analyses of the coating were measured in each one of three samples.

Kumar et al. [16] obtained a porosity of 0.98% for an 86WC-10Co4Cr coating using HVOF deposition; likewise (Guaglianoni et al. [80] obtained porosities of 0.5 and 1% in WC–Co-Cr coatings using the same technique HVOF. On the other hand, Thakur et al. [81] compared two WC–Co-Cr coatings obtained by the HVOF with

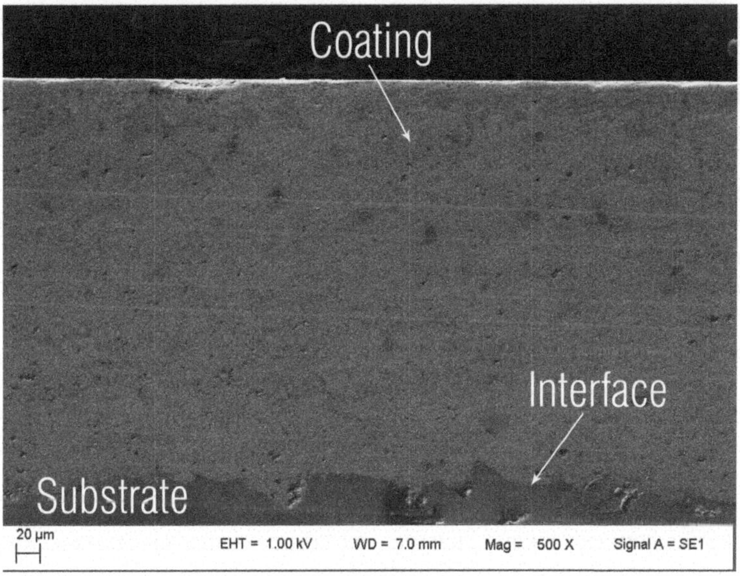

Fig. 10 SEM image of the microstructure of the HVOF 86WC-10Co-4Cr coating in the cross-Section (500×). Reproduced with permission from [26]

different grain size carbides (100, 200, and 500 nm), obtaining porosities of 0.75% and 0.57%, respectively.

The microhardness values along the cross-section of the sprayed coating often undergo variations due to the non-uniformity of the phases. According to Castro et al. [78], it is possible to observe that when working with sublayers (thermal spray coatings), there is a variation in microhardness values, as each indentation point can be located in different microstructures, in this case, carbides, oxides, inclusions and the matrix itself. As described by Solution et al. [82] WC-based materials are presented as metal-ceramic composites, and the microhardness values found are associated with each micro constituent. The microstructure depends on the composition of each phase, the morphology, and size of the sprayed powder, the spraying technique, porosity, among other aspects.

Santacruz et al. [26] reported the highest value of microhardness performed on the 86WC-10Co-4Cr HVOF coating was 1448 HV0.3, being also the closest one to the manufacturer's specification Oerlikon Metco Company [75]. Thakur et al. [81] and Castro et al. [83] obtained similar results of 1297 ± 45 HV0.3 and 1256 HV, respectively, for WC-CoCr coatings sprayed by HVOF. In the same way, Kumar et al. [16], Maiti et al. [31], Berger et al. [84], and Ahmed et al. [17] reported microhardness values of 1031 ± 99, 1180 ± 70, 1118 ± 131, and 1106 HV0.3 for similar sprayed coatings applied by the same technique, respectively.

4 Erosion in CERMET Coatings

Erosion in cermet is very complex due to its non-homogeneous character. However, the classical theory of erosion referring to ductile or brittle material nature cannot be used to fully explain their erosion.

According to Hussainova et al. [85], the erosion of WC–Co cermets is associated with the combination of ductility and brittle behavior in erosion, although brittle is dominant. This has to do with the different behavior of the different classes of materials exposed to erosion; for example, concerning the impact angle, at low angles, ceramic materials reduce erosive wear due to their high hardness and toughness [86]. On the other hand, if the material is subject to the impact of particles at an angle close to the normal (90°), the exposed surface must be able to withstand repeated deformations. In this case, metallic materials are preferred over ceramics, in which cracks progress rapidly and lead to the removal of the material. This relationship between material properties and wear resistance is shown in Fig. 11 [76].

However, in some situations, there is a mixture of mechanisms that lead to erosive wear, such as cutting, chipping, fatigue, and brittle fracture mechanisms. In these cases, there is a controversy between the hardness and fracture toughness of materials and coatings. According to Kulu et al. [76], metallic matrix composite coatings reinforced with ceramic particles partially solve this problem, as shown in Fig. 11. Coatings with higher hardness and lower tenacity are better suited to resist wear at attack angles of eroding particles up to 30°.

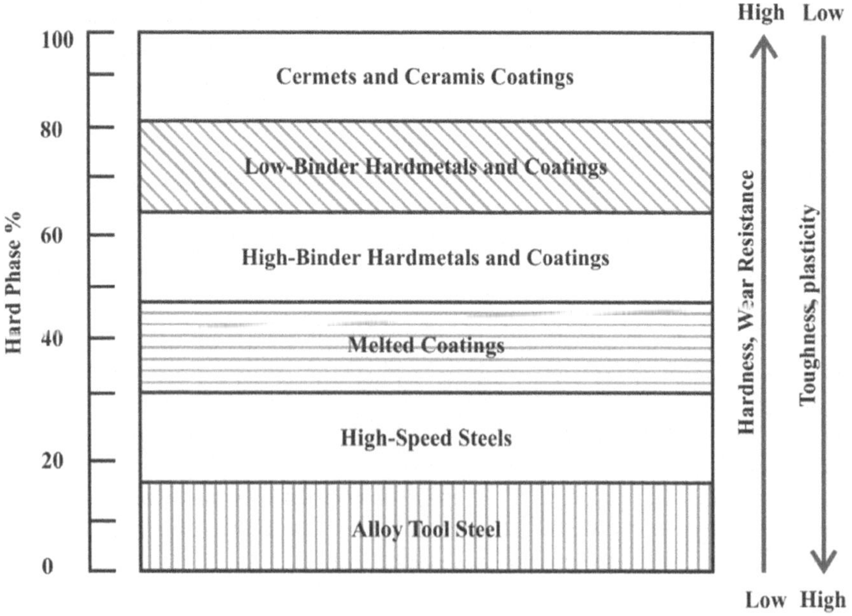

Fig. 11 Wear resistance materials and coatings. Edited and reproduced with permission from [76]

Thermally sprayed materials, such as cermets, are often used to resist different forms of wear, in various industrial applications, in aqueous media, and even in situations of extreme erosion. In wear conditions where there is a mixture of mechanisms, the thermally sprayed coatings based on WC–Co are highly effective [76].

WC grains are hard and resistant enough to dissipate a large amount of energy in the impact of eroding particles without microcracking the grains. Therefore, according to Engqvist et al. [87] and Hussainova et al. [85], the mechanical properties and wear resistance of cermet depend on each grain, the matrix, the connection between the phases, and the adjacent grains.

The combination of physical and mechanical properties of ceramics with metals, such as the high hardness, mechanical strength, rigidity, and wear resistance of cermets, make them widely used as coating materials in the hydrometallurgical industry [88, 89].

4.1 *Wear Definition in Materials*

Wear and tear have been known since human beings started using natural elements that served as household items. This phenomenon, such as corrosion and fatigue, is one of the most important forms of degradation of parts, mechanical elements, and industrial equipment.

According to ASTM G40-17 [90], wear can be defined as altering a solid surface by the progressive loss or progressive displacement of material due to the relative movement between this surface and the substance in contact. This phenomenon is usually manifested on the surfaces of materials, thus affecting their microstructure and, therefore, the loss of properties. In many cases, wear is initiated by one mechanism and can be continued by other wear mechanisms.

The wear for engineering purposes is described, according to Bayer et al. [91], as being surface damage caused by mechanical interaction with another surface, body, or fluid. In this interaction, the so-called wear mechanisms would act, which involve a series of physical and chemical phenomena, and become worrying as the damage caused interferes with the proper functioning of the component in question.

The damage mechanisms in materials are mainly due to plastic deformation, crack formation and propagation, corrosion, and/or wear. There is no consensus when it comes to defining and classifying the many wear processes. Each wear process involves phenomena that characterize it. In many situations, one of the wear modes acts so that it influences the others [60, 92].

According to DIN 50 320 [93], there are eight wear mechanisms: adhesion, abrasion, erosion, and tribochemical reaction (corrosion-wear). However, this same standard also mentions that, in addition to these mechanisms, there are other types of wear, such as cavitation, contact fatigue, fretting, and scuffing, which are also causes of material deterioration.

According to ASTM G40-17 [90], abrasive wear is the loss of mass resulting from the interaction between hard particles or roughness forced against a surface and moves along with it. The difference between abrasive wear and slip or adhesive wear is the degree of wear between the bodies involved (greater in abrasive wear), whether by nature, material type, chemical composition, or geometrical configuration.

The wear rate depends on the degree of penetration of the abrasive into the material's surface being abraded. The wear will be greater the harder the abrasive is concerning the surface that is suffering wear [94]. The worn surface is subject to a very high load (in a small area) which not only causes the penetration of the abrasive but can also cause the fracture of hard phases present on the surface and the plastic deformation of the matrix. According to Dong et al. [95], the factors that affect the abrasive wear rate are the surface and abrasive properties, as well as the nature and severity of the interactions between them.

Adhesive wear is associated with any formation and subsequent breakage of adhesive bonds between interfaces when two surfaces are placed in contact and material transfers from one surface to another during relative movement due to the formation of joints in the solid phase [94]. The origin of this type of wear is generally based on the same phenomenon that is responsible for friction.

Cavitation wear is a phenomenon where small bubbles are formed in a fluid subjected to high speeds or high energy sound waves. This phenomenon wears out the material due to the large transient pressures during bubble collapse. Microbubbles are formed in low-pressure regions during the flow and collapse in high-pressure regions, removing material from the surface and forming microcavities [96].

This type of wear is less aggressive than erosion and appears to be controlled by the fatigue strength of the materials. The occurrence of this type of wear is common in most machines subject to vibrations during operation, such as hydraulic machines, piping, and in general, in any device where there is fluid flow/flow [97, 97].

4.2 *Erosion Definition in Materials*

According to ASTM G76-13 [98], erosion is defined as the progressive loss of material from a solid surface due to the mechanical interaction between the surface and a fluid, a multicomponent fluid, or impacting liquid or solid particles.

The erosion phenomenon has been studied since the nineteenth century, but the first technical article on erosion appeared at the beginning of the twentieth century [86]. This means that erosion in materials has been seen for a long time in many technological and engineering systems. However, a better-founded analysis of the erosion process occurred only later, involving analyzing the structure of eroded surfaces [99].

According to Bhushan et al. [100], erosion is a type of wear caused by the repetitive impact of solid or liquid particles or the implosion of bubbles formed in the fluid against the surface of a solid body.

Erosion is often observed in a wide variety of environmental conditions, especially those involving fluid transport (liquids or gases), with or without the presence of particulates. As examples of equipment and components that are subject to this type

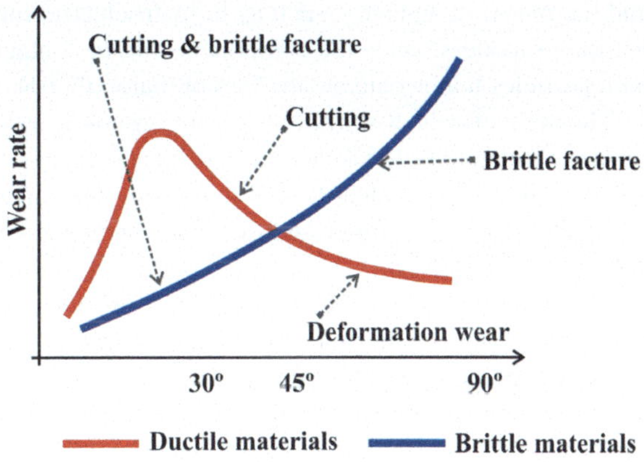

Fig. 12 Erosive wear rate as a function of particle incidence angle (collision) for ductile and brittle materials

of wear, the following are cited: pneumatic conveyors, turbines, pipes, pumps, valves, connections, among others.

Erosion can be classified as a function of six different erosion wear mechanisms: (a) erosion by solid particles in a gaseous medium; (b) erosion by solid particles in a liquid medium; (c) erosion due to the action of liquids; (d) erosion-corrosion, (e) erosion by cavitation and (f) thermal erosion. Each type of erosion has its particular mechanism and depends on specific conditions.

According to Berthier et al. [101], in ductile materials, such as metals and polymers, erosive wear occurs preferentially through plastic deformation, displacement, or cutting action of the eroding particle.

In fragile materials, such as ceramics, are highly susceptible to cracks and microcracks. Therefore, they are removed by the interconnection of cracks that diverge from the point of impact of the eroding particle with the surface [102].

The velocity of the particles, the angle of impact, and the size of the abrasive particles provide a proportion of the kinetic energy of the particles' collision. Wear particles are formed in erosion as a result of successive impacts. The relationship between wear rate and impact angle is different for ductile and brittle materials, as shown in Fig. 12. For ductile materials, the maximum erosion occurs at approximately 30° angles, and in the case of brittle materials, the maximum erosion rate occurs in normal impacts. The shape of the abrasive particles also affects the wear rate. Flattened particles lead to a more localized deformation and have higher wear rates than rounded particles [103].

4.3 *Slurry Erosion in Materials*

The degradation of material caused by impacts from a solid particle (erodent) suspended in a liquid is called slurry erosion. Degradation caused by slurry occurs in fluid flow machines, mainly in the hydroelectric industry, the mining industry, and the maritime industry, such as in hydroelectric turbines, valves, pipelines, and marine propellers, since the components are in contact with the liquid and contain hard particles that accelerate the loss of material [104, 105].

The main effects of slurry erosion are economic losses resulting from the need to replace or regenerate the device and stop the technological process. In addition, slurry erosion causes the gradual loss of mass from the surface layer, alters the geometry of the elements of facilities and machines, which leads to a reduction in efficiency and useful life [106].

The failure costs associated with slurry erosion are due to part replacements or regeneration of corroded machine parts, lost productivity, and indirect energy losses, becoming an industrial, financial problem.

Since 1946, when the first technical works on erosive wear were published by Finnie et al. [86], many investigations have been carried out on material degradation by slurry. In 1960 Finnie et al. [107] published their pioneering work about slurry erosion, in which degradation mechanisms and theoretical analyzes to predict erosion damage was presented.

Many researchers have carried out additional work with theoretical considerations and extensive experimental investigations, but slurry-like erosion still remains an unresolved wear problem.

According to Buszko et al. [108], the number of works dedicated to the slurry erosion of solid particles indicates the degree of these problems. Only in 2020, in Elsevier's publications, more than 700 articles, which concerned the phenomenon of slurry-like erosion, were published.

All works were developed trying to better understand the phenomenon of slurry-like erosion and the correlations between material properties and material strength, using various types of test equipment simulating different conditions.

4.4 Main Factors Responsible for the Slurry Erosion Process in Materials

Degradation of materials due to slurry erosion depends on many factors, which can be divided into three main groups: the first linked to fluid flow conditions (flow velocity, particle incidence angle, particle concentration, liquid density, liquid chemical activity, liquid temperature), the second has to do with the solid particles (size, shape, hardness, strength) and the third with the target material (mechanical and strength properties: toughness, fatigue, yield and maximum strength, work hardening, surface topography, microstructure, number and size of defects) [105, 108, 109]. Thus, the number of factors influencing slurry erosion is vast, and the degradation of materials is a synergistic effect of all the mentioned factors.

In Fig. 13, we can see a slurry erosion scheme which is a process of surface degradation due to repeated impacts of solid particles flowing with a liquid.

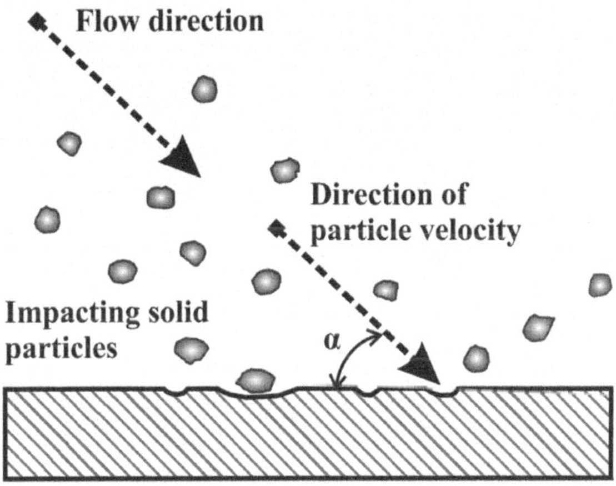

Fig. 13 Schematic of slurry erosion

Impact velocity and solid particle size influence the kinetic impact energy of a single particle. As the impact velocity and particle size increase, the erosion rate increases. A similar effect on the erosion rate has an increase in the concentration of particles in a liquid, which influences the total impact energy [108].

The dynamic fluid forces acting on the eroding particles due to flow patterns over the target also interfere with the actual impact angle. When the jet of fluid impacts the flat surface, the jet will spread across the surface. In slurry erosion, the direction of eroding will be the flow direction due to the high viscosity of the liquid [110].

According to Grewal et al. [104], slurry erosion is classified according to impact velocity as high and low-velocity erosion. If the velocity is between 6 and 9 m/s, erosion is classified as high velocity, while below this velocity, erosion is called low velocity erosion. The presence of sub micrometric particles (30–150 μm) in diameter) in a liquid that flows with a velocity above 10 m/s develops a subsonic erosion [111].

The next key factor that has a considerable impact on this kind of erosion is the hardness of the material. As the hardness of the target material increases, the critical incidence angle at which the maximum erosion rate occurs increases, as shown in Fig. 12. In the case of ductile materials, the maximum erosion rate occurs at approximately 30◦. In the case of brittle or hard materials, they reach the maximum erosion rate at the impact angle of roughly 90° under normal incidence. Thus, despite the hardness of the material, the hardness of solid particles influences the erosion rate of the material [112, 113].

4.5 *Erosion Mechanisms*

Three erosion mechanisms are very important and often identified in the removal of material under the erosion of solid particles, which are micro-cut (dominant

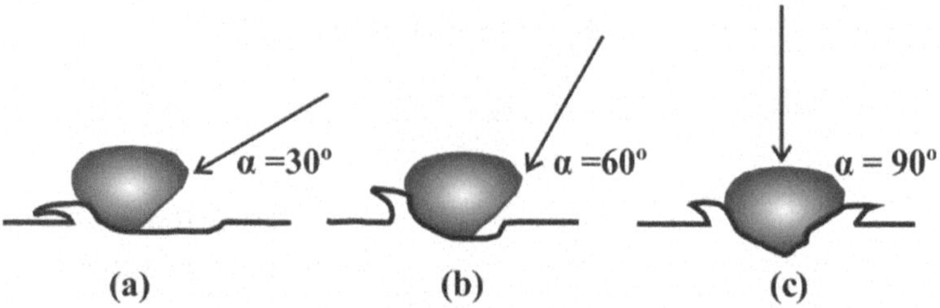

Fig. 14 The effect of incidence angles on the mode of material removal

for high angle erosion of ductile materials), micro-plow (prevalent for low angle erosion of materials ductile), and microfracture (dominant for brittle materials [15, 86, 114, 115].

The theory of solid particle erosion assumes that, for small incidence angles of approximately 30°, the erosion is due to a chip formation process resulting from the micro-plowing action of the erodent particles.

How eroded particles remove material from a ductile target surface involves displacement of the material, producing craters. An impact particle pushes the material forward, leaving a groove in its wake and a lip at the end of the groove as it leaves the incidence site. This is shown schematically in Fig. 14a.

For high incidence angle values of approximately 90°, the erosion mechanism is no longer a cut and can be described as repeated impact wear. In ductile materials, plastic deformation usually results from particles impacting the surface, as shown in Fig. 14c. When the impact energy exceeds the limit, under steady-state conditions, the surface layer is strongly deformed, resulting in ductile microfractures.

For incidence angles between 30° and 90°, both mechanisms contribute to the erosion process, as shown in Fig. 14b.

A jet slurry erosion test was carried out by Santacruz et al. [25] in a modified commercial high-pressure washer based on the ASTM G-76 standard [98] detailed described it. This equipment allows the control of the incidence angle, the incidence velocity, the concentration of erodent particles in the suspension, and the test temperature [110, 116–118]. Figure 15 shows the configuration of the testing advice [25].

Table 3 the accumulated jet slurry erosion rate by incidence erodent mass, in volume loss of the samples, as a function of incidence angle, in a total test time of 4 min, for the materials used in the research of martensitic stainless steel AISI 410 and Tungsten Carbide (86WC-10Co4Cr) HVOF coating.

Table 3 shows the accumulated erosion rate by mass of erodent in loss of volume of the samples tested, as a function of the impact angle, in a total time of 4 min, martensitic stainless steel AISI 410 and Tungsten Carbide (86WC-10Co4Cr) HVOF coating [25].

In tests carried out with incidence angles of 30° and 90°, the 86WC-10Co4Cr HVOF coating had the lowest jet slurry erosion than martensitic stainless steel AISI 410.

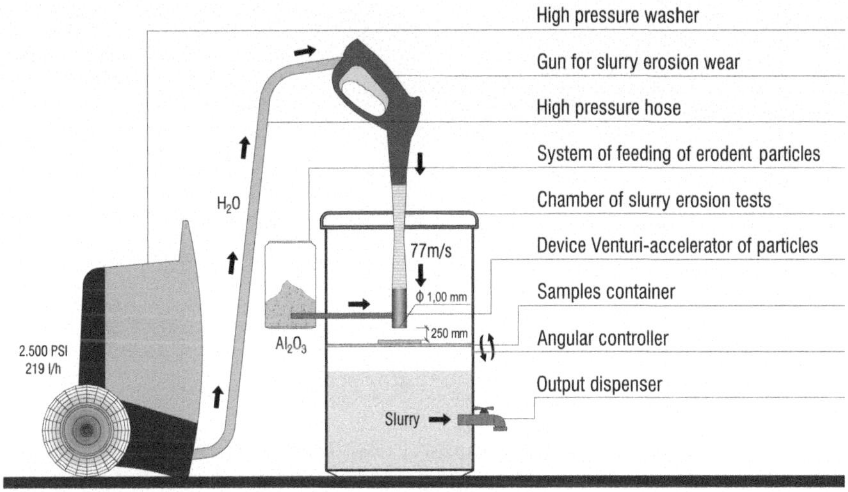

Fig. 15 Schematic diagram of the jet slurry erosion tester. Reproduced with permission from [25]

Table 3 Accumulated erosion rate in volume loss [25]

Materials/coating	Impact angle (°)	Volumetric erosion rate $(cm^3{}_{target}/g_{erodent}) * 10^{-5}$	Standart deviation $(cm^3{}_{target}/g_{erodent}) * 10^{-5}$
Steel AISI 410	30	0.8449	0.0114
	90	0.7025	0.0495
86WC-10Co4Cr	30	0.4490	0.0023
	90	0.6778	0.0283

The researchers concluded that the 30° angles of incidence of the 86WC10Co4Cr tungsten carbide coating had similar performance and accumulated volumetric erosion rate, approximately 50% lower than the martensitic stainless steel AISI 410. For the 90° incidences angles, the 86WC10Co4Cr tungsten carbide coating gets a slightly lower cumulative volumetric wear rate than AISI 410 martensitic stainless steel, as shown in Table 3 and Fig. 16 [25].

Figure 16 [25] shows that the AISI 410 martensitic stainless steel showed higher erosive wear at an incidence angle of 30° than at 90°, with ductile material behavior similar to that reported by the authors Islam et al. [119] and Okonkwo et al. [120].

According to Santacruz et al. [25], the material removal mechanisms for martensitic stainless steel AISI 410 at an angle of 30° is by plowing, through successive plastic deformations until the material is removed. The study agrees with the results of several authors that confirm that metallic coatings have a higher wear rate at lower angles [91, 121–123].

On the other hand, Santacruz et al. [25] concluded that the tungsten carbide coating (86WC-10Co4Cr) presented higher erosive wear at 90° incidence angles than at 30°, indicating a fragile behavior of this coating. This is due to the repeated

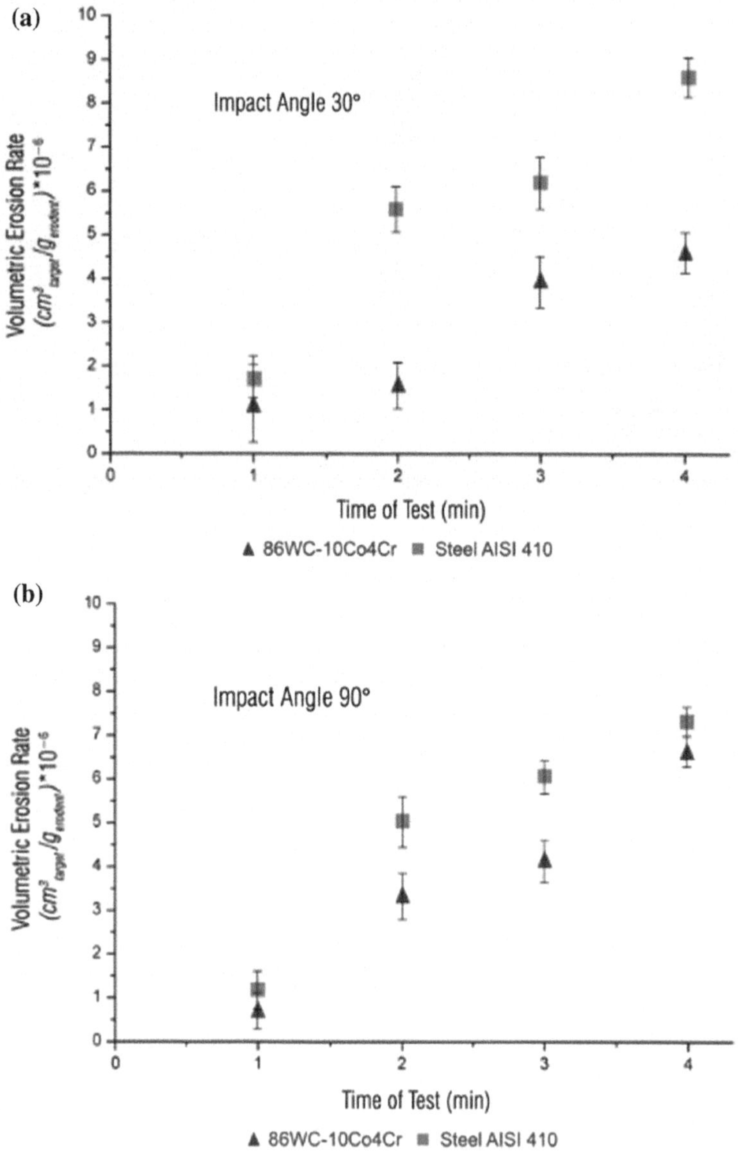

Fig. 16 Variation of the accumulated volumetric erosion rate as a function of the incidence angle of 30° and 90° for the martensitic stainless steel (AISI 410) and the 86WC10Co4Cr tungsten carbide coating. Reproduced with permission from [25]

action of erosive particles perpendicular to the surface, producing microfractures that contribute to the wear of the carbide WC, resulting in its eventual removal, in agreement with the results of Barber et al. [124].

References

1. Popov, et al.: Adhesive wear: Generalized rabinowicz'criteria. Facta Universitatis, Ser. Mech. Eng. **16**(1), 29–39 (2018)
2. Suyan, et al.: Enhanced wear resistance of iron-based alloy coating induced by ultrasonic impact. Coatings **9**(12), 804 (2019)
3. Dauber et al.: Erosion resistance of engineering ceramics and comparative assessment through wiederhorn and evans equations. Wear **432**, 202938 (2019)
4. Abe et al.: Precipitate design for creep strengthening of 9% cr tempered martensitic steel for ultra-supercritical power plants. Sci. Technol. Adv. Mater. **9**(1), 013002 (2008)
5. Henao et al.: Principles and applications of thermal spray coatings. In: Advanced Surface Coating Techniques for Modern Industrial Applications, pp. 31–70. IGI Global (2021)
6. Fragassa et al.: Advances in Design by Metallic Materials: Synthesis, Characterization, Simulation and Applications (2021)
7. López, et al.: Evaluation of protective coatings for high-corrosivity category atmospheres in offshore applications. Materials **12**(8), 1325 (2019)
8. Huang et al.: Slurry erosion behaviour and mechanism of hvof sprayed micro-nano structured wc-cocr coatings in nacl medium. Tribol. Int. **148**, 106315 (2020)
9. Wang, et al.: Effect of wc grain size and abrasive type on the wear performance of hvof-sprayed wc-20cr3c2-7ni coatings. Coatings **10**(7), 660 (2020)
10. Mohrbacher et al.: Martensitic automotive steel sheet-fundamentals and metallurgical optimization strategies. In: Advanced Materials Research, vol. 1063, pp. 130–142. Trans Tech Publ (2015)
11. Klueh, et al.: Ferritic/martensitic steels for next-generation reactors. J. Nucl. Mater. **371**(1–3), 37–52 (2007)
12. Babic et al.: New method for modeling the topographical property of metals and its application in robot laser hardening with overlapping. In: IOP Conference Series: Materials Science and Engineering, vol. 659, pp. 012084. IOP Publishing (2019)
13. Toro, et al.: Corrosion–erosion of nitrogen bearing martensitic stainless steels in seawater–quartz slurry. Wear **251**(1), 1257–1264 (2001)
14. Krauss et al.: Steels: Processing, Structure, and Performance. Asm International (2015)
15. Levy, et al.: Erosion of steels in liquid slurries. Wear **98**, 163–182 (1984)
16. Kumar, et al.: Effect of spray particle velocity on cavitation erosion resistance characteristics of HVOF and HVAF processed 86wc-10co4cr hydro turbine coatings. J. Therm. Spray Technol. **25**(6), 1217–1230 (2016)
17. Ahmed, et al.: Sliding wear investigation of suspension sprayed wc–co nanocomposite coatings. Wear **322**, 133–150 (2015)
18. Rolison, et al.: Nanoparticles: synthesis. Properties and Applications. Institute of Physics Publishing, London, UK (1996)
19. Scrivani, et al.: A contribution to the surface analysis and characterisation of HVOF coatings for petrochemical application. Wear **250**(1–12), 107–113 (2001)
20. Bolelli, et al.: A comparison between the corrosion resistances of some HVOF-sprayed metal alloy coatings. Surf. Coat. Technol. **202**(19), 4793–4809 (2008)
21. Camargo et al.: Resistance to abrasive and erosive wear of WCCo/NiCr HVOF coatings: Comparative evaluation of commercial nanostructured spray powders

22. Bergmann et al.: Protection against erosive wear using thermal sprayed cermet: a review. In: Protection against Erosive Wear Using Thermal Sprayed Cermet, pp. 1–77. Springer, Berlin (2011)
23. Gunn et al.: Duplex Stainless Steels: Microstructure, Properties and Applications. Woodhead Publishing (1997)
24. Peckner et al.: Handbook of Stainless Steels. McGraw-Hill Book Co., New York (1977) (Chapters paged separately) (1977)
25. Santacruz, et al.: Comparative study of jet slurry erosion of martensitic stainless steel with tungsten carbide HVOF coating. Metals **9**(5), 600 (2019)
26. Santacruz et al.: Experimental analysis of jet slurry erosion on martensitic stainless steel. In: IOP Conference Series: Materials Science and Engineering, vol. 659, pp. 012077. IOP Publishing (2019)
27. Davis et al.: Alloy Digest Sourcebook: Stainless Steels. ASM international (2000)
28. Garverick et al.: Corrosion in the Petrochemical Industry. ASM international (1994)
29. Carrouge et al.: Microstructural change in high temperature heat-affected zone of low carbon weldable '13 % cr' martensitic stainless steels. Stain. Steel World (The Netherlands) **14**,16–17 (2002)
30. Jaber, et al.: Dissimilar spot welding of dual phase steel/ferritic stainless steel: phase transformations. Sci. Technol. Weld. Joining **19**(7), 565–571 (2014)
31. Maiti, et al.: Improving the wear behavior of WC-CoCr-based HVOF coating by surface grinding. J. Mater. Eng. Perform. **18**(8), 1060 (2009)
32. Oyama et al.: Introduction to the chemistry of transition metal carbides and nitrides. In: The Chemistry of Transition Metal Carbides and Nitrides, pp. 1–27. Springer, Berlin (1996)
33. Pierson et al.: Handbook of Refractory Carbides and Nitrides: Properties, Characteristics, Processing and Applications. William Andrew (1996)
34. Garcia, et al.: Cemented carbide microstructures: a review. Int. J. Refract Metal Hard Mater. **80**, 40–68 (2019)
35. Raymond et al.: Modern Physical Metallurgy and Materials Engineering. Butterworth-Heinemann (1999)
36. Xie, et al.: Comparative investigation on HVOF sprayed carbide-based coatings. Appl. Surf. Sci. **273**, 799–805 (2013)
37. Schwetzke, et al.: Microstructure and properties of tungsten carbide coatings sprayed with various high-velocity oxygen fuel spray systems. J. Therm. Spray Technol. **8**(3), 433–439 (1999)
38. Chivavibul, et al.: Effects of carbide size and co content on the microstructure and mechanical properties of HVOF-sprayed WC–Co coatings. Surf. Coat. Technol. **202**(3), 509–521 (2007)
39. Jafari, et al.: Comparison between oxidation kinetics of HVOF sprayed WC–12Co and WC–10co–4Cr coatings. Int. J. Refract Metal Hard Mater. **41**, 78–84 (2013)
40. Edelstein et al.: Nanomaterials: Synthesis, Properties and Applications. CRC press (1998)
41. Matteazzi, et al.: Synthesis of nanocrystalline alumina–metal composites by room- temperature ball-milling of metal oxides and aluminum. J. Am. Ceram. Soc. **75**(10), 2749–2755 (1992)
42. Suryanarayana, et al.: Synthesis of nanocomposites by mechanical alloying. J. Alloy. Compd. **509**, S229–S234 (2011)
43. Schacht et al.: Refractories Handbook, vol. 178. CRC Press (2004)
44. Guaglianoni, et al.: Influence of milling parameters in the morphology of nanostructured CrC-30wt% NiCr. Tecno-Lógica **21**(1), 37–40 (2017)
45. Sherif, et al.: Synthesis and characterizations of ball-milled nanocrystalline WC and nanocomposite WC–Co powders and subsequent consolidations. J. Alloy. Compd. **312**(1–2), 315–325 (2000)
46. Zhang, et al.: Parameters optimization in the planetary ball milling of nanostructured tungsten carbide/cobalt powder. Int. J. Refract Metal Hard Mater. **26**(4), 329–333 (2008)
47. Hewitt, et al.: Effects of ball milling time on the synthesis and consolidation of nanostructured WC–Co composites. Int. J. Refract Metal Hard Mater. **27**(6), 937–948 (2009)

48. Mahmoodan, et al.: Microstructural and mechanical characterization of high energy ball milled and sintered WC–10 wt% Co–xTaC nano powders. Int. J. Refract Metal Hard Mater. **27**(4), 801–805 (2009)
49. Enayati, et al.: Production of nanostructured WC–Co powder by ball milling. Int. J. Refract Metal Hard Mater. **27**(1), 159–163 (2009)
50. Raihanuzzaman, et al.: Characterization of short-duration high-energy ball milled WC–Co powders and subsequent consolidations. J. Alloy. Compd. **615**, S564–S568 (2014)
51. Mandel, et al.: Particle properties of submicron-sized WC–12Co processed by planetary ball milling. Int. J. Refract Metal Hard Mater. **42**, 200–204 (2014)
52. Suryanarayana, et al.: Mechanical alloying and milling. Prog. Mater Sci. **46**(1–2), 1–184 (2001)
53. Suryanarayana, et al.: Mechanically alloyed nanocomposites. Prog. Mater Sci. **58**(4), 383–502 (2013)
54. Koch, et al.: Processing of metals and alloys. Mater. Sci. Technol. Comprehensive Treatment **15**, 193–245 (1991)
55. Atzmon, et al.: In situ thermal observation of explosive compound-formation reaction during mechanical alloying. Phys. Rev. Lett. **64**(4), 487 (1990)
56. Gavrilov et al.: Computer simulation of mechanical alloying in a shaker ball mill. In: Tenth International Conference on Composite Materials. III. Processing and Manufacturing (1995)
57. Dhiman, et al.: Predicting splat morphology in a thermal spray process. Surf. Coat. Technol. **201**(18), 7789–7801 (2007)
58. Zhang et al.: Nanostructured Thin Films and Coatings: Mechanical Properties. CRC Press (2010)
59. Zhang et al.: Studies of splat morphology and rapid solidification during thermal spraying. Int. J. Heat Mass Transfer **44**(24), 4579–4592 (2001)
60. Davis et al.: Handbook of Thermal Spray Technology. ASM International (2004)
61. Zheng et al.: Modeling the impact, flattening and solidification of a molten droplet on a solid substrate during plasma spraying. Appl. Surf. Sci. **317**, 526–533 (2014)
62. Alavi, et al.: Simulation of semi-molten particle impacts including heat transfer and phase change. J. Therm. Spray Technol. **21**(6), 1278–1293 (2012)
63. Cinca, et al.: An overview of intermetallics research and application: Status of thermal spray coatings. J. Mater. Res. Technol. **2**(1), 75–86 (2013)
64. Fincke, et al.: Diagnostics and control in the thermal spray process. Surf. Coat. Technol. **146**, 537–543 (2001)
65. Planche, et al.: Relationships between nicrbsi particle characteristics and corresponding coating properties using different thermal spraying processes. Surf. Coat. Technol. **200**(7), 2465–2473 (2005)
66. Acuña et al.: Evaluation of tribological wear and corrosion in coatings of diamalloy 4060ns deposited by thermal spray. J. Phys. Conf. Ser. **935**, 012018 (2017). IOP Publishing
67. Nautiyal et al.: High-temperature erosive wear behavior of high-velocity oxy-fuel sprayed Cr3C225 (Ni20Cr) coating on (AISI 316) austenitic steel. J. Tribol. **142**(7) (2020)
68. Vernhes, et al.: HVOF coating case study for power plant process control ball valve application. J. Therm. Spray Technol. **22**(7), 1184–1192 (2013)
69. Trevisan et al.: Aspersao termica fundamentos e aplicacoes. Sao Paulo, Brasil: Editorial Artliber (2002)
70. Tipton et al.: The Effect of HVOF Thermal Spray on the Elevated Temperature High Cycle Fatigue Behavior of a Martensitic Stainless Steel. Dresser-Rand, New York (2002)
71. Li et al.: Effect of gas conditions on HVOF flame and properties of WC-Co coatings. Mater. Manuf. Processes **14**(3), 383–395 (1999)
72. Clark, et al.: Wear rates and specific energies of some ceramic, cermet and metallic coatings determined in the coriolis erosion tester. Wear **233**, 319–327 (1999)
73. Cabrera et al.: Design and Development of a High Velocity Oxy-Fuel Thermal Spray Gun. The University of Texas at El Paso (2014)

74. Santacruz et al.: Estudo comparativo do desgaste erosivo do tipo lama, do aço inox martensitíco AISI 410 com os revestimentos carbeto de tungstênio 86WC-10Co4Cr, obtido por HVOF e o elastômero PUR comercial. Master's Thesis, Federal University of Rio Grande do Sul, Porto Alegre, Brazil (2018)
75. Oerlikon Metco. Oerlikon Metco Company (2021)
76. Kulu, et al.: Solid particle erosion of thermal sprayed coatings. Wear **258**(1–4), 488–496 (2005)
77. Rayes, et al.: Erosion-corrosion of cermet coating. Int. J. Electrochem. Sci. **8**(1), 1117–1137 (2013)
78. Castro et al.: Microestructura de materiales compuestos WC-Co/Ni-W-Cr recubiertos por HVOF. Revista Latinoamericana de Metalurgia y Materiales **21**(1) (2001)
79. ASTM Standard. E2109 01 (2014) standard Test Methods for Determining Area Percentage Porosity in Thermal Sprayed Coatings. ASTM International, Pennsylvania
80. Guaglianoni et al.: Síntese, caracterização e aplicação por HVOF de um nanocompósito de WCCoCr/NiCr para uso como revestimento protetor contra o desgaste erosivo em equipamentos da indústria do petróleo (2015)
81. Thakur, et al.: An investigation on erosion behavior of HVOF sprayed WC–CoCr coatings. Appl. Surf. Sci. **258**(3), 1225–1234 (2011)
82. Trelleborg Sealing Solution. Hidraulic seals/rod seals. Catálogo de Produtos, Suécia (2011)
83. Castro et al.: Avaliação das características de superfície e do desgaste abrasivo de revestimentos aplicados em hastes de cilindros hidráulicos pelas técnicas de aspersão (HVOF) e eletrodeposição. PhD thesis. Federal University of Rio Grande do Sul, Porto Alegre, Brazil (2012)
84. Berger, et al.: Microstructure and properties of HVOF-sprayed chromium alloyed WC–Co and WC–Ni coatings. Surf. Coat. Technol. **202**(18), 4417–4421 (2008)
85. Hussainova, et al.: Mechanical properties and features of erosion of cermets. Wear **250**(1–12), 818–825 (2001)
86. Finnie, et al.: Some reflections on the past and future of erosion. Wear **186**, 1–10 (1995)
87. Engqvist, et al.: Tribological properties of a binderless carbide. Wear **232**(2), 157–162 (1999)
88. Wang, et al.: Elevated temperature erosion of HVOF Cr3C2/TiC–NiCrMo cermet coating. Wear **233**, 342–351 (1999)
89. Upadhyaya, et al.: Materials science of cemented carbides—an overview. Mater. Des. **22**(6), 483–489 (2001)
90. ASTM Standard. Astm g40–17 (2017) Standard Terminology Relating to Wear and Erosion. ASTM International, Pennsylvania
91. Bayer et al.: Mechanical wear prediction and prevention. Marcel Dekker, Inc, P. O. Box 5005, Monticello, NY 12701–5185, USA, 1994. 657 (1994)
92. Handbook, A.S.M.: Friction, lubrication and wear technology. ASM Int. **18**, 362–369 (1992)
93. DIN Norma. 50320: análise sistemática dos processos de desgaste. Classificação dos fenômenos de desgaste. Metalurgia e Materiais **53**, 619–622 (1997)
94. Ludema et al.: Friction, Wear, Lubrication. A Textbook in Tribology. CRC Press (1996)
95. Dong et al.: Tribological characteristics of α-alumina at elevated temperatures. J. Am. Ceramic Soc. **74**(5), 1036–1044 (1991)
96. Harvey, et al.: On cavity formation in water. J. Appl. Phys. **18**(2), 162–172 (1947)
97. Szeri et al.: Tribology: Friction, Lubrication, and Wear. McGraw Hill (1980)
98. ASTM Standard.: Astm g76–13 (2013) standard test method for conducting erosion tests by solid particle impingement using gas jets. ASTM International, Pennsylvania (2013)
99. Shewmon, et al.: The erosion of metals. Annu. Rev. Mater. Sci. **13**(1), 301–318 (1983)
100. Bhushan et al.: Principles and Applications of Tribology. Wiley, New York (2013)
101. Berthier, et al.: Experimental evidence for friction and wear modelling. Wear **139**(1), 77–92 (1990)
102. Hoppert, et al.: Alumina ceramics–superior materials for protection against wear and corrosion. Aufbereit.-Tech. **30**(10), 625–634 (1989)

103. Ingole et al.: Tribology for Scientists and Engineers: From Basics to Advanced Concepts. Springer, Berlin (2013)
104. Grewal, et al.: Design and development of high-velocity slurry erosion test rig using CFD. J. Mater. Eng. Perform. **22**(1), 152–161 (2013)
105. Grewal, et al.: Identifying erosion mechanism: A novel approach. Tribol. Lett. **51**(1), 1–7 (2013)
106. Grewal, et al.: Slurry erosion performance of Ni-Al_2O_3 based thermal-sprayed coatings: effect of angle of impingement. J. Therm. Spray Technol. **23**(3), 389–401 (2014)
107. Finnie et al.: Erosion of Surfaces by Solid Particles. Wear **3**(2), 87–103 (1960)
108. Buszko, et al.: Slurry erosion–design of test devices. Adv. Mater. Sci. **17**(2), 5–17 (2017)
109. Lathabai, et al.: Microstructural influence in slurry erosion of ceramics. Wear **189**(1–2), 122–135 (1995)
110. Zu, et al.: Design of a slurry erosion test rig. Wear **140**(2), 331–344 (1990)
111. Arora et al.: Cavitation inception on microparticles: A self-propelled particle accelerator. Phys. Rev. Lett. **92**(17), 174501 (2004)
112. Shitole, et al.: Effect of impacting particle kinetic energy on slurry erosion wear. J. Bio-Tribo-Corrosion **1**(4), 29 (2015)
113. Oka, et al.: Practical estimation of erosion damage caused by solid particle impact: Part 1: Effects of impact parameters on a predictive equation. Wear **259**(1), 95–101 (2005)
114. Hutchings, et al.: A model for the erosion of metals by spherical particles at normal incidence. Wear **70**(3), 269–281 (1981)
115. Ball, et al.: The mechanisms of wear, and the performance of engineering materials. J. South Afr. Inst. Min. Metall. **86**(1), 1–13 (1986)
116. Wentzel, et al.: The erosion-corrosion resistance of tungsten-carbide hard metals. Int. J. Refract Metal Hard Mater. **15**(1–3), 81–87 (1997)
117. Santa, et al.: Slurry erosion of thermal spray coatings and stainless steels for hydraulic machinery. Wear **263**(1), 258–264 (2007)
118. Santa, et al.: Slurry and cavitation erosion resistance of thermal spray coatings. Wear **267**(1), 160–167 (2009)
119. Islam, et al.: Effect of impact angle and velocity on erosion of API X42 pipeline steel under high abrasive feed rate. Wear **311**(1–2), 180–190 (2014)
120. Okonkwo, et al.: Erosion behaviour of API X100 pipeline steel at various impact angles and particle speeds. Metals **6**(10), 232 (2016)
121. Papini, et al.: The plowing erosion of organic coatings by spherical particles. Wear **222**(1), 38–48 (1998)
122. O'Flynn et al.: A model to predict the solid particle erosion rate of metals and its assessment using heat-treated steels. Wear **248**(1–2), 162–177 (2001)
123. Biswas et al.: Erosion wear behavior of polymer composites: a review. J. Reinforced Plast. Compos. **29**(19), 2898–2924 (2010)
124. Barber, J., Mellor, B.G., Wood, R.J.K.: The development of sub-surface damage during high energy solid particle erosion of a thermally sprayed WC–Co–Cr coating. Wear **259**(1–6), 125–134 (2005)

Nanostructured Thermoelectric Materials

Janio Venturini

Abstract This chapter discusses the recent achievements in the performance of nanostructured thermoelectric devices. The main mechanisms behind the thermoelectric effect are introduced, and the figure-of-merit of this phenomenon is presented and its main components. Selected nanoscale approaches to decreasing lattice thermal conductivity are discussed. The influence of low-dimensionality and nanoinclusions on the thermoelectric performance of these devices is demonstrated. Techniques for mesoscale engineering are presented, with a particular focus on spark plasma sintering. Finally, the advantages of hierarchical architectures are discussed.

Keywords Thermoelectric materials · Nanostructuring · Hierarchical approach

Abbreviations

HIP	Hot isostatic pressuring
mfp	Mean free path
MS-SPS	Melt-spinning spark plasma sintering
SALT	Sodium antimony lead telluride
SPS	Spark plasma sintering

1 Introduction

Our unprecedented level of technological development coupled with an ever-growing world population poses a significant problem to our energy supplies. Currently, our energy matrix is firmly based on three fossil fuels: oil, natural gas, and coal. Besides these sources being major contributors to greenhouse gas emissions, their utilization is also jarringly inefficient. Most of the energy contained in these fuels is lost in

J. Venturini (✉)
UFRGS - Universidade Federal do Rio Grande do Sul, Porto Alegre, Brazil

A. Kopp Alves (ed.), *Technological Applications of Nanomaterials*,
Engineering Materials, https://doi.org/10.1007/978-3-030-86901-4_2

the form of heat. For instance, the internal combustion engine of our current cars converts only about 20% of the energy contained in the fuel into useful work [1]. Over half of the total energy content is lost in the form of heat and friction.

One of the forms of minimizing this energy waste is the utilization of thermoelectric materials. These semiconductors are able to transform a temperature gradient into a voltage difference. In other words, they can convert heat waste into nobler, more useful forms of energy. Therefore, their utilization may help recover a larger fraction of the energy contained in the system, thus increasing the overall efficiency of the process.

When heat is applied to one end of a material, the charge carriers in this region become more energetic. The higher kinetic energy means these carriers migrate towards the colder side more rapidly than colder electrons in the opposite direction. Also, the electronic distribution at the hot side is smeared out, with more electrons above the Fermi level. On the other hand, the cold side shows a much sharper distribution. The net effect is the development of a charge movement towards the colder side in an attempt to lower the energy of the system. When both sides of the material are connected in a circuit, a potential difference is observed. The development of a voltage under these conditions is the so-called Seebeck effect. A representation of the movement of charge carriers along a temperature gradient is shown in Fig. 1.

The relation between temperature gradient ΔT and generated voltage V_{oc} is given by the Seebeck coefficient (S), which is defined in Eq. 1.

$$S = -\frac{V_{oc}}{\Delta T} \tag{1}$$

The Seebeck coefficient usually has a magnitude of a few microvolts per Kelvin. Nickel and gold at room temperature, for instance, display Seebeck coefficients of −18 and + 1.8 μV/K, respectively [2]. These low values mean that very little potential is generated, even with steep temperature gradients. The most recent advanced modules reach maximum conversion efficiencies of close to 7% [3]. Although already pushing the limits of the field, these values are much lower than those of current generators,

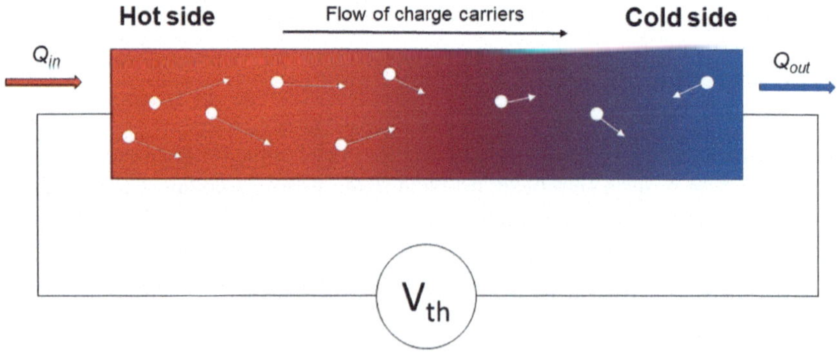

Fig. 1 Representation of the Seebeck effect in a doped semiconductor

such as water and gas turbines. Therefore, thermoelectric generators are usually relegated to very niche applications, such as in space probes [4]. Nevertheless, increases in the efficiency of these green devices could lead to their increased utilization. While it is clear that thermoelectric generators will not supply a large fraction of our energy needs, they could tap into an enormous amount of waste heat generated during energy conversion, be it in a power plant or an automobile.

2 Figure of Merit

The performance of a thermoelectric device is generally described by a dimensionless figure of merit (ZT). This value is defined as a function of Seebeck coefficient S, electrical conductivity σ, temperature T, and thermal conductivity κ, as described in Eq. 2. The ZT values for most commonly utilized thermoelectric devices hover just below 1, with efficiencies between 5 and 8%.

$$ZT = \frac{S^2 \sigma T}{\kappa} \tag{2}$$

The thermal conductivity may be further separated into two contributions: the heat transported by electron flow, κ_e, and the heat transported by the lattice via phonons, κ_l.

$$ZT = \frac{S^2 \sigma T}{\kappa_e + \kappa_l} \tag{3}$$

Any attempt at improving the efficiency of future thermoelectric materials should focus on improving the ZT value. Materials with a low number of charge carriers usually lead to maximized Seebeck coefficient. On the other hand, the electrical conductivity σ increases in the other direction. Therefore, a compromise is usually reached via the utilization of heavily-doped semiconductors, which maximize the $S^2\sigma$ relation, also known as the power factor. The optimization of the components of ZT as a function of the charge carrier concentration is displayed in Fig. 2.

As the power factor is reasonably optimized in a given material, the usual strategy for improving the ZT of a device involves tinkering with the denominator of Eq. 3. The heat transferred by electronic movement, κ_e, is strongly linked to the electronic conductivity; any decrease in κ_e is accompanied by a reduction in σ, rendering any efforts in this direction counterproductive. Therefore, the thermal contribution from the lattice κ_l is the factor that could be decreased to good effect, as it is the only component that is reasonably decoupled from the other variables that define the figure of merit.

Indeed, methods that affect the mean free path (mfp) of phonons without considerably leading to losses in electronic conductivity have recently produced materials with ZT of over 2. Some of the more recent ZT results of selected selenides and

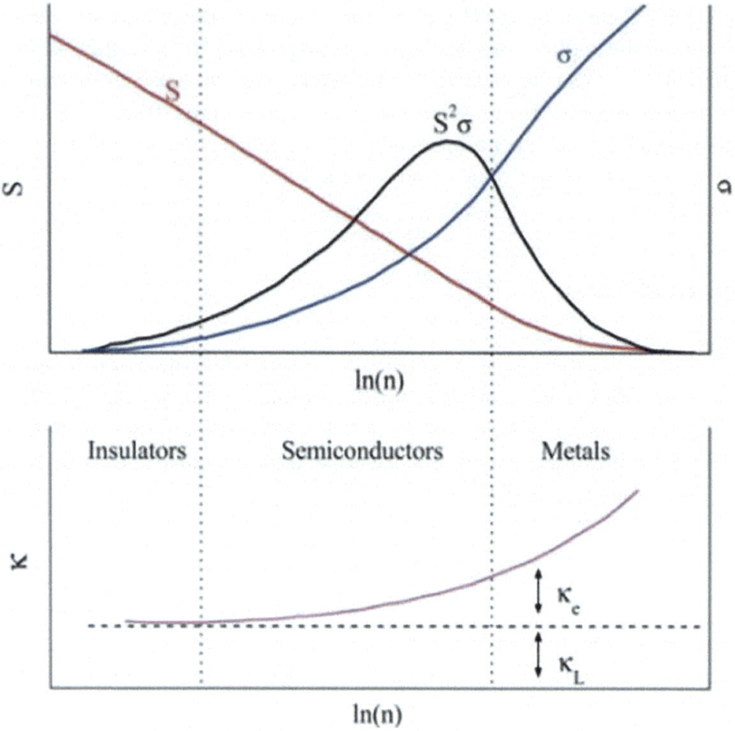

Fig. 2 Behavior of the components of ZT as a function of the number of charge carriers. Reprinted from [5] with permission from The Royal Society of Chemistry

tellurides can be seen in Fig. 3, adapted from a work by Liu and collaborators [6]. Most of these successful approaches are linked to the utilization of some degree of nanoengineering of thermoelectric devices.

3 Nanostructured Approaches

In the 90s, new strategies of phonon structure engineering shook the thermoelectric materials out of their relative stagnation. Hicks and Dresselhaus were the leading proponents behind this novel approach, which was based on enhanced phonon scattering at the boundaries of nanostructured materials [7, 8]. In a nutshell, if the characteristic dimensions of the structures are larger than the mfp of the electrons but shorter than the mfp of phonons, thermal conductivity may be reduced without significant effects on the power factor ($S^2\sigma$).

Silicon nanowires were utilized in some of the first attempts at exploiting low dimensionality. Though the sharper band structure of 1D materials may directly lead

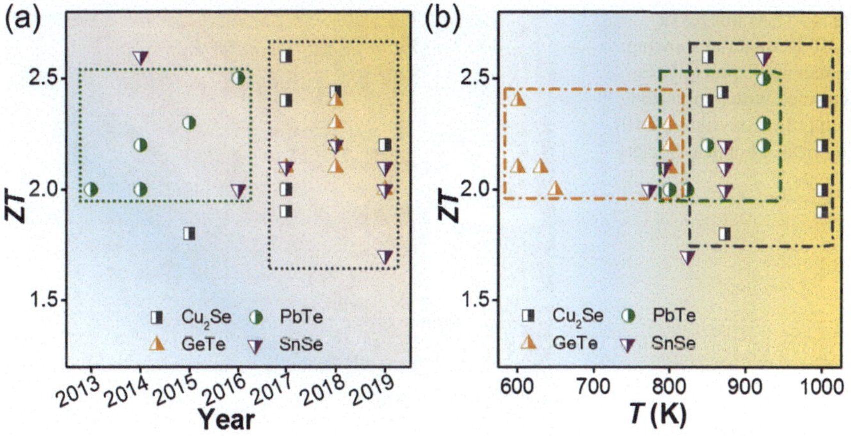

Fig. 3 Recently reported ZT values of Cu_2Se, SnSe, PbTe, and GeTe by **a** year of publication and **b** temperature at which the maximum ZT is achieved. Reprinted from [6] with permission from Elsevier

to improvements in power factor, the increased probability of phonon scattering at the surfaces is also conducive to enhanced ZT values. Silicon nanowires have shown ZT values of up to 0.7 at 1200 K [9], several times larger than that of bulk silicon (~0.01). Silicon–germanium nanowires may show even higher figures of merit. An n-type SiGe alloy has shown a peak in ZT of 1.3 at 1173 K [10]. Simulations show that the localized constriction of these nanowires may help enhance the thermoelectric performance depending on the constriction width [11]. A macroscopic all-silicon thermoelectric generator built of nanowire forests has shown a remarkable output power density of approximately 1 $\mu W/(cm^2.K^2)$ [12].

Another successful strategy in the production of effective thermoelectric nanowires is the utilization of core–shell morphologies. Yang and collaborators [13] have recently shown that the utilization of different materials may lead to efficient decoupling of phonon and electron transport. In the report, a tellurium wire covered with a PEDOT:PPS shell shows a ZT value of 0.54 at 400 K. The authors propose that the electrical transport occurs through the organic shell. In contrast, thermal conduction occurs mainly via the metallic wire, thus separating these two components of the thermoelectric performance.

Chen et al. subdivided the complex effects of the nanowire morphology on the thermal conduction mechanisms into five phenomena: classic size effect, acoustic softening, complex morphology, dimensional crossover, and surface roughness. The latter has been shown to play quite an essential role in the screening of phonons [5]. Simulations show a $(D/\Delta)^2$ dependence of thermal conductivity on both the diameter D and the surface roughness Δ of a nanowire [14]. For instance, smooth ($\Delta \sim 0.2$ nm) 115 nm Si nanowires show thermal conductivities five times higher than their rough ($\Delta \sim 3$ nm) analogs, 40 and 8 W/mK at 200 K, respectively.

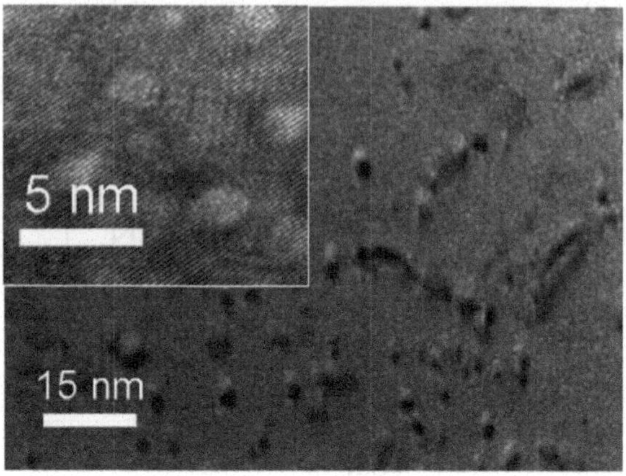

Fig. 4 TEM images of SALT-18 material showing the nanoscale precipitates. Reprinted with permission from [15]. Copyright (2010) American Chemical Society

The other nanostructured approach commonly utilized for bulk thermoelectrics is the creation of interfaces with a characteristic size comparable to that of the phonons. One such method is via the precipitation of nanoparticles. Materials of the SALT system (sodium antimony lead telluride, $NaPb_mSbTe_{2+m}$) naturally show elevated figures of merit, with some compositions reaching values of close to 2. Kanatzidis has shown that this eccentricity is accounted for by the presence of nanoinclusions in the matrix [15]. This material is not composed of a single solid solution but of nanoclusters embedded in a matrix, as shown in Fig. 4. During cooling, phase segregation occurs, creating clusters rich in Na and Sb embedded in a lead telluride matrix. Similar behavior is observed in related materials, such as the LAST family (lead antimony silver telluride, $AgPb_mSbTe_{2+m}$).

When the m value of the LAST series is above 10, the material presents itself as a single phase in XRD. However, these compounds also show the nanoinclusions present in the SALT materials. The random distribution of Na^+ and Sb^{3+} required in solid solutions is disturbed by Coulombic forces, which lead the system towards clustering at the nanoscale. The generated structures promote increased phonon scattering at the interfaces, which in turn translates into the elevated ZT values of these families of materials. For instance, the decreased thermal conductivity of LAST-18 ($AgPb_{18}SbTe_{20}$) results in ZT values of close to 2 at 900 K [16].

Besides nanoinclusions, interfaces may be engineered into the material via careful processing. Interfaces decrease thermal conductivity by scattering phonons with mfp larger than the distance between interfaces. Therefore, the careful design of materials with a high density of interface area per unit volume should lead to decreases in the lattice thermal contribution. However, the creation of bulk thermoelectrics requires a heat treatment for the consolidation of the grains, such as hot pressing. These processing techniques generally lead to recrystallization, with the growth of nanosized grains and consequent loss of much of the interfacial area between the grains. Nevertheless, one particular approach is gaining ground in the processing of thermoelectrics.

Spark plasma sintering (SPS) is a method for consolidating precursors based on the passage of a current through a powder compact. When a spark discharge happens between two neighboring particles, the local temperature increases violently, causing melting or evaporation of the materials and forming necks between the particles. The reduced sintering times associated with this technique lead to consolidation of the compact without excessive grain growth. Therefore, SPS allows the fabrication of dense ceramics with very fine grains in shorter times. The large volumetric fraction of grain boundaries associated with SPS processing is very interesting for thermoelectric devices, as the increased density of interfaces should translate into decreased thermal conduction due to phonon scattering.

Delaizir et al. [17] have studied the effects of different types of sintering on the thermoelectric performance of p-type Bi_2Te_3. The results indicate that SPS leads to an increase in ZT of approximately 30% (0.53 to 0.68) compared to hot isostatic pressuring (HIP). Among others, the authors relate the improvement in performance to the smaller grain size of the ceramics produced via SPS. Studies on nanograined bismuth telluride alloys also report an improvement in electrical properties when SPS is utilized as a consolidation method instead of HIP [18]. However, the complex effects of the processing technique may also lead to decreased power factors, as reported for ZnO-based materials [19]. Nevertheless, SPS has been successfully used to produce several thermoelectric devices of ZT close to 2 [20], the value above which these materials are expected to be technically viable.

The SPS technique may also lead to the production of finer inclusions when compared to the conventional method. This effect was demonstrated in higher manganese silicides (HMS), a family of thermoelectric materials of great promise due to the wide availability of its components [21]. The ZT of the studied HMS was increased almost 100% (to 0.62) with the processing via melt-spinning SPS (MS-SPS). When prepared via conventional methods, MnSi inclusions are generally present in these materials in the form of micrometric grains. Processing via conventional SPS leads to increased ZT, albeit without affecting the morphology of the inclusions. On the other hand, the treatment via MS-SPS markedly decreased the size of these inclusions to a few tens of nanometers. The smaller size of the inclusions translated into improved phonon scattering with the consequent increased thermoelectric performance. The morphologies created by the different processing conditions are exhibited in Fig. 5.

4 Hierarchical Approach

Despite the great success of the presented nanoscale approaches, a thermoelectric device with a viable ZT value (above 2) still cannot be realized by these methods alone. The main issue with these strategies lies in the fact that they usually only affect a small subset of the lattice phonons. For instance, while nanoinclusions may effectively scatter short mfp phonons, those with mid and long mfp are still able to conduct heat through the material. An example of the contribution of each type

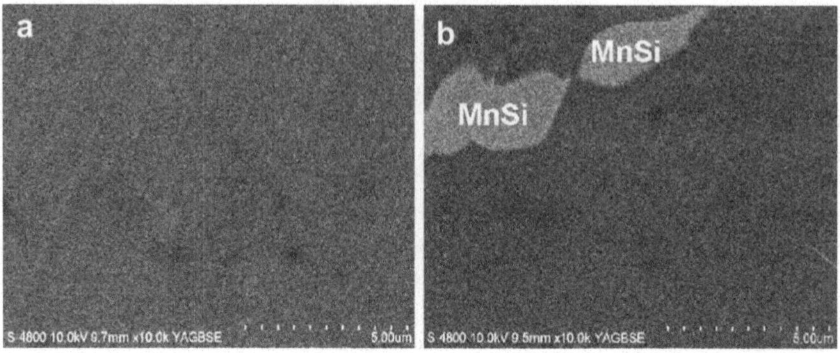

Fig. 5 BSE images of bulk higher manganese silicide (HMS) processed via **a** melt-spinning SPS and **b** conventional method. Reprinted from [21] with permission from Elsevier

of phonon towards the total lattice thermal conductivity is shown in Fig. 6. As can be perceived in the spectra, there is a wide distribution of thermal transport along phonons of different mfps. In this regard, Zhao, Dravid, and Kanatzidis proposed a hierarchical structuring of thermoelectric devices, with the design of varying levels of organization in the system, aiming to simultaneously dissipate phonons along the entire spectrum [22].

Point defects and nanoinclusions produce strong scattering of short and mid wavelength phonons, thus denying their contribution towards thermal conductivity. Nevertheless, long wavelengths are still able to transfer heat along with the extent of the material. Additional levels of structural engineering are needed to scatter these carriers. Albeit the definition of 'long' varies depending on the material, the scattering of these phonons is usually achieved by structuring the crystals in the mesoscale, generally from 100 nm to 5 μm [22].

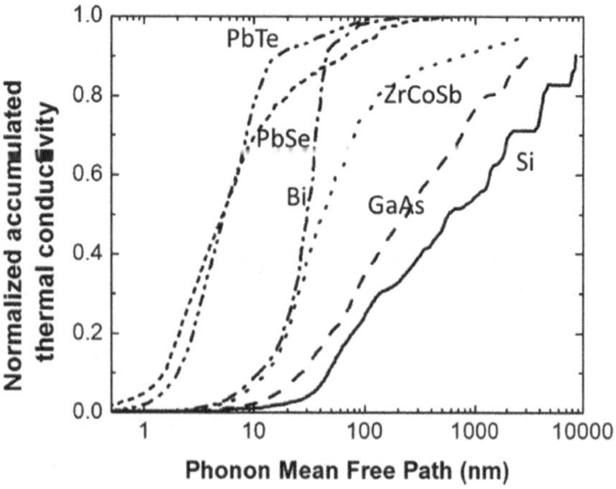

Fig. 6 Calculated contribution of phonons of different mfp towards thermal conductivity of crystalline materials. Reprinted from [23] with permission of the American Chemical Society

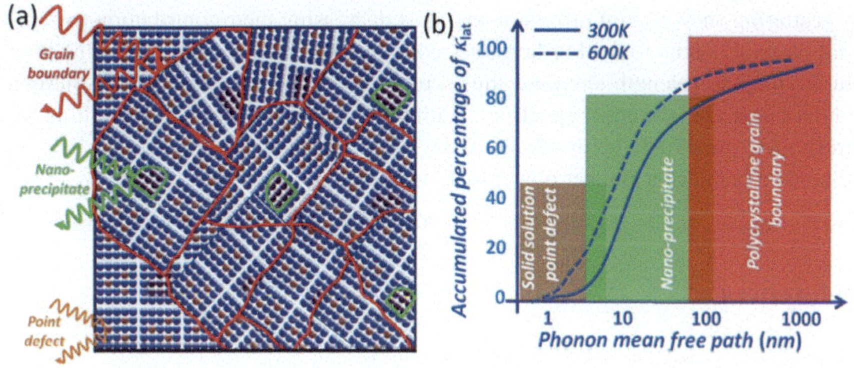

Fig. 7 **a** Different scales of a hierarchical architecture and **b** cumulative contribution to the thermal conductivity of phonons of different mean free paths in Si or PbTe bulk. Adapted from [22] with permission from The Royal Society of Chemistry

The panoscopic approach entails utilizing an all-level hierarchical structure, tailoring the thermoelectric material at different scales to achieve maximal efficiency. At an atomic level, point defects and 'rattler' atoms are used in order to scatter shorter wavelengths. At larger dimensions, nanoprecipitates and nanostructuring are applied to screen the heat transport of mid wavelength phonons. Finally, mesostructured grains effectively filter out the contribution of long mfp phonons towards total heat conductivity. An example of the hierarchical architecture and the filtering strategies is shown in Fig. 7. The effects of this complex design are leading to significant gains in the performance of thermoelectric devices.

A very instructive example of the power of the hierarchical approach can be seen in the work of Biswas and collaborators [24]. The authors present an optimized Na-doped PbTe structure with a maximum ZT of 1.1. The addition of SrTe nanoinclusions leads to an increase in ZT to approximately 1.7, owing to the efficient scattering of phonons with short and mid mfp. However, the authors go beyond mere nanostructuring. Instead of melt-processing, the final step in their hierarchical strategy involved processing via SPS in order to better control the mesostructure of the material. The special processing led to a 30–50% increase in ZT, with a maximum of 2.2 at 915 K. The authors relate the excellent thermoelectric performance to the integrated scattering of phonons across all length scales, which is reflected in the minimized thermal conductivity of the material.

5 Closing Remarks

The topics discussed in this chapter demonstrated the many approaches that are currently being studied for the production of ever more efficient thermoelectric devices. A nanostructured design may help increase the thermoelectric figure of merit by scattering short to mid mfp phonons, thus decreasing their contribution towards total thermal conductivity. Furthermore, a hierarchical architecture also helps filter out the long-wavelength phonons, thus screening the lattice component of thermal conduction across the entire spectrum. With these strategies, we are approaching the threshold above which thermoelectrical devices may become commonplace, helping humankind achieve a greener power matrix.

References

1. Rodriguez, R., Preindl, M., Cotton, J.S., Emadi, A.: Review and trends of thermoelectric generator heat recovery in automotive applications. IEEE Trans. Veh. Technol. **68**, 5366–5378 (2019). https://doi.org/10.1109/TVT.2019.2908150
2. Macia, E.: Thermoelectric Materials: Advances and Applications, 1st edn. Pan Stanford (2015)
3. Liu, Z., Sato, N., Gao, W. et al.: Demonstration of ultrahigh thermoelectric efficiency of ~7.3% in Mg3Sb2/MgAgSb module for low-temperature energy harvesting. Joule **5**, 1196–1208 (2021). https://doi.org/10.1016/j.joule.2021.03.017
4. Beekman, M., Ghantous, J.F., Thomson, K.: Potential error from using ZT to optimize thermoelectric performance. AIP Adv. **11**, 55207 (2021). https://doi.org/10.1063/5.0048411
5. Vaqueiro, P., Powell, A.V.: Recent developments in nanostructured materials for high-performance thermoelectrics. J. Mater. Chem. **20**, 9577–9584 (2010). https://doi.org/10.1039/C0JM01193B
6. Liu, W.-D., Yang, L., Chen, Z.-G.: Cu2Se thermoelectrics: property, methodology, and device. Nano Today **35**, 100938 (2020). https://doi.org/10.1016/j.nantod.2020.100938
7. Hicks, L.D., Dresselhaus, M.S.: Effect of quantum-well structures on the thermoelectric figure of merit. Phys. Rev. B **47**, 12727–12731 (1993). https://doi.org/10.1103/PhysRevB.47.12727
8. Hicks, L.D., Dresselhaus, M.S.: Thermoelectric figure of merit of a one-dimensional conductor. Phys. Rev. B **47**, 16631–16634 (1993). https://doi.org/10.1103/PhysRevB.47.16631
9. Bux, S.K., Blair, R.G., Gogna, P.K. et al.: Nanostructured bulk silicon as an effective thermoelectric material. Adv. Funct. Mater. **19**, 2445–2452 (2009). https://doi.org/10.1002/adfm.200900250
10. Wang, X.W., Lee, H., Lan, Y.C., et al.: Enhanced thermoelectric figure of merit in nanostructured n-type silicon germanium bulk alloy. Appl. Phys. Lett. **93**, 193121 (2008). https://doi.org/10.1063/1.3027060
11. Zianni, X.: Monte Carlo simulations on the thermoelectric efficiency of a si nanowire modulated by a constriction. Mater Today Proc. **3**, 840–846 (2016). https://doi.org/10.1016/j.matpr.2016.02.017
12. Elyamny, S., Dimaggio, E., Magagna, S., et al.: High power thermoelectric generator based on vertical silicon nanowires. Nano Lett. **20**, 4748–4753 (2020). https://doi.org/10.1021/acs.nanolett.0c00227
13. Yang, L., Gordon, M.P., Menon, A.K. et al.: Decoupling electron and phonon transport in single-nanowire hybrid materials for high-performance thermoelectrics. Sci. Adv. **7**, eabe6000 (2021). https://doi.org/10.1126/sciadv.abe6000
14. Martin, P., Aksamija, Z., Pop, E., Ravaioli, U.: Impact of phonon-surface roughness scattering on thermal conductivity of thin Si nanowires. Phys. Rev. Lett **102**, 125503 (2009). https://doi.org/10.1103/PhysRevLett.102.125503
15. Kanatzidis, M.G.: Nanostructured Thermoelectrics: The New Paradigm? Chem. Mater. **22**, 648–659 (2010). https://doi.org/10.1021/cm902195j
16. Hsu, K.F., Loo, S., Guo, F. et al.: Cubic $AgPb_mSbTe_{2+m}$: bulk thermoelectric materials with high figure of merit. Science (80-) **303**, 818 LP–821 (2004). https://doi.org/10.1126/science.1092963

17. Delaizir, G., Bernard-Granger, G., Monnier, J. et al.: A comparative study of Spark Plasma Sintering (SPS), Hot Isostatic Pressing (HIP) and microwaves sintering techniques on p-type Bi_2Te_3 thermoelectric properties. Mater. Res. Bull. **47**, 1954–1960 (2012). https://doi.org/10.1016/j.materresbull.2012.04.019
18. Virta, J., Tervo, J.: Experimenting with hot isostatically pressed (HIP) nano grained bismuth-telluride-based alloys. AIP Conf. Proc. **1449**, 544–547 (2012). https://doi.org/10.1063/1.4731614
19. Tian, T., Cheng, L., Xing, J. et al.: Effects of sintering on the microstructure and electrical properties of ZnO-based thermoelectric materials. Mater. Des. **132**, 479–485 (2017). https://doi.org/10.1016/j.matdes.2017.07.033
20. Srinivasan, B., Gellé, A., Gucci, F., et al.: Realizing a stable high thermoelectric zT$\sim$2 over a broad temperature range in $Ge_{1-x-y}Ga_xSb_yTe$ via band engineering and hybrid flash-SPS processing. Inorg. Chem. Front. **6**, 63–73 (2019). https://doi.org/10.1039/C8QI00703A
21. Luo, W., Li, H., Yan, Y. et al.: Rapid synthesis of high thermoelectric performance higher manganese silicide with in-situ formed nano-phase of MnSi. Intermetallics **19**, 404–408 (2011). https://doi.org/10.1016/j.intermet.2010.11.008
22. Zhao, L.-D., Dravid, V.P., Kanatzidis, M.G.: The panoscopic approach to high performance thermoelectrics. Energy Environ. Sci. **7**, 251–268 (2014). https://doi.org/10.1039/C3EE43099E
23. Lee, S., Chen, G.: Nanostructured thermoelectric materials. In: Innovative Thermoelectric Materials, pp. 77–105. Imperial College Press (2014)
24. Biswas, K., He, J., Blum, I.D., et al.: High-performance bulk thermoelectrics with all-scale hierarchical architectures. Nature **489**, 414–418 (2012). https://doi.org/10.1038/nature11439

Nanomaterials for Viral Detection

Pedro Augusto Machado Vitor and Gabriela Machado Parreira

Abstract The mass testing tactic is among the main strategies to fight a virus pandemic. It allows for an early diagnosis in the initial phase of the disease and reduces disease transmission. In this sense, there is a growing interest in developing devices with high sensitivity, selectivity, and fast detections. With this purpose, nanobiosensors are presented as a promising alternative, produced from nanomaterials with different structures and properties. On biosensing, NMs comprise transduction elements (transducers) associated with biomarkers to recognize and amplify different signals when interacting with biological material. The primary transducers involve optical and electrochemical methods. Gold nanoparticles (AuNPs) and carbon-based, such as graphene, graphene oxide, and carbon nanotube (CNT), make up most NMs used in biosensing. For such application, the use of magnetic nanoparticles (MNPs) and quantum dots (QDs) of different compositions, such as the basis of cadmium and tellurium (CdTe QDs), are also widely studied. In addition to applications in biosensing, nanomaterials can be applied in biomarker immobilization and extraction procedure in standard tests such as RT-PCR and LFIA (ELISA). NMs allow for the improvement of different techniques used in viral detection, presenting diverse and unique solutions for health crisis moments, including for Covid-19.

Keywords Nanomaterials · Virus detection · Nanobiosensor · Covid-19

Abbreviations

AIDS	Acquired immunodeficiency syndrome
CA	Chronoamperometry
CDs	Carbon-dots
CQDs	Carbon quantum-dots

P. A. M. Vitor (✉)
Universidade Federal do Rio Grande do Sul, Porto Alegre, Brazil

G. M. Parreira
Universidade Federal de Minas Gerais, Belo Horizonte, Brazil

A. Kopp Alves (ed.), *Technological Applications of Nanomaterials*,
Engineering Materials, https://doi.org/10.1007/978-3-030-86901-4_3

CoV	Coronaviruses
CoV-2	Coronaviruses 2
CV	Cyclic voltammetry
DENV	Dengue virus
DPV	Differential pulse voltammetry
EBOV	Ebola virus
ECL	Electrochemiluminescence
EIS	Electrochemical impedance spectroscopy
ELISA	Enzyme-linked immunosorbent assay
FAdVs	Fowl adenoviruses
FET	Field-effect transistor
FL	Fluorescence
FSNIA	Fluorescent silver nanoparticle-based immunoassay
GO	Graphene oxide
GQDs	Graphene quantum dots
H1N1	Influenza A virus subtype
H5N1	Influenza A virus subtype
HBV	Hepatitis B virus
HCV	Hepatitis C virus
HIV	Human Immunodeficiency Virus
HIV-1	Human Immunodeficiency Virus subtype 1
HIV-2	Human Immunodeficiency Virus subtype 2
HPV	Human papilloma virus
LFIA	Lateral flow immunoassay
LSPR	Localized surface plasmon resonance
LSV	Linear sweep voltammetry
MERS-CoV	Middle East respiratory syndrome coronavirus
MWCNTs	Multi-walled carbon nanotubes
NCs	Nanoclusters
NMs	Nanomaterials
NPs	Nanoparticles
NRs	Nanorods
NTs	Nanotubes
MNPs	Magnetic nanoparticles
PPT	Plasmonic photo-thermal
QDs	Quantum dots
RGO	Reduced graphene oxide
RT-PCR-real-time	Polymerase chain reaction
SARS-CoV-2	Severe acute respiratory syndrome coronavirus 2
SPR	Surface plasmon resonance
SWV	Square wave voltammetry

1 Introduction

Our history is marked by numerous outbreaks of viral infections that were responsible for a large number of deaths. For example, in 2015, the Americas witnessed an epidemic caused by the Zika virus; in 2003, there was an outbreak of severe acute respiratory syndrome (SARS), Africa witnessed recurrent crises between 2013 and 2016 caused by the EBOLA virus, and, since 2019, the Covid-19 (coronavirus disease) has been able to infect more than 200 million and kill more than 4 million people, in the worst global health crisis in recent times.

The virus has a simple structure consisting of a layer of protein, nucleic acid, and viral enzymes, which may or may not have a lipid envelope [1]. Furthermore, viruses use the host's cellular machinery for replication and are therefore intracellular pathogens. Viral infections can affect many organs and cause several diseases, such as hepatitis B virus (HBV), affecting the liver, influenza virus, and coronavirus, affecting the respiratory tract and lungs; the gastrointestinal tract can be affected by rotavirus, among others.

One of the first steps in remediating a virus is detecting its presence. The sooner it is detected, the greater the chances of success. Therefore, fast and reliable diagnostics are needed to control future pandemics. There are currently different viral detection methods that involve molecular biology and serological methods, known as the RT-PCR and ELISA tests, respectively. Usually, these methods are time-consuming, laborious, and often expensive [2].

Herefore, further studies are needed to discover new methodologies and technologies. In this context, detection techniques from biosensors have been presented as an alternative. This biosensing method combines a biologic material (a bioreceptor) with a physical transductor, detecting and amplifying a signal from different events in a possible biological interaction, as in optical, electrochemical, and mass change events [3]. In the development of physical transductors, nanomaterials (NMs) and nanotechnology assume an essential role in sensing detection.

Due to their exceptional electronic and surface properties, the nanomaterials have been fundamental to developing and improving different viral detection methods, which has enabled an accurate, selective, and primarily fast diagnostic [4]. Besides biosensing, the NMs may also act in immobilization, isolation, and extraction of biomarkers in conventional diagnostics, as immunoassays (ELISA) [4].

Several studies using different NMs have shown promise, which the most used nanomaterials are nanoparticles (NPs) of gold (AuNPs), graphene oxides (GO, RGO), and quantum dots (QDs). The electrical, mechanical, and functional characteristics of NMs, enabling the construction of different systems that act on improvement from detection to application in biomedicine. According to the Science Direct database, a summary of the papers developed in the last ten years offering some coverage in the area of virus detection through these main nanomaterials is shown in Fig. 1. There is a significant increase in publications related to the use of NPs, especially gold (AuNPs) and quantum dots of different compositions, to viral detection techniques. Also, there is a growing interest in carbon-based nanostructures, such as

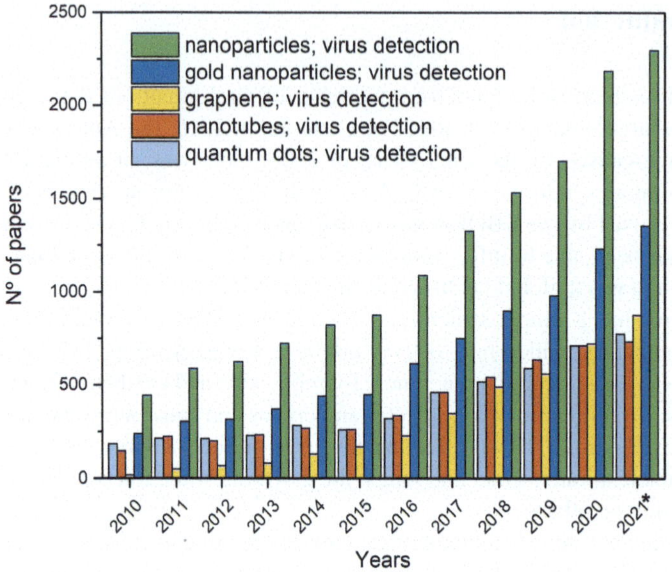

Fig. 1 Rate of scientific papers published between 2010 and 2021* in the Science Direct database, using the term "virus detection" associated with the words "nanoparticles," "gold nanoparticles," "graphene," "nanotubes," and "quantum dots." *Survey carried out in June 2021

carbon nanotubes (CNTs) and the promising graphene and their oxides, in detection strategies of different viruses.

This prospective chapter reports the current strategies based on nanotechnology and the NMs developed and applied to optimize the detection of different viruses such as Sars-CoV-2 (COVID-19), H1N1 (Influenza A), HBV-HCV (Hepatitis B and C), HIV (AIDS), among others.

2 Strategies and Techniques for Viral Detection

Currently, different techniques are used to detect viruses, and each one is used at different stages of the infection. First, there is a latency period in which it is not yet possible to detect the immune system's response, so techniques are used to detect the presence of the virus (direct detection). The methods are based on molecular biology methods, the RT-PCR method (real-time -polymerase chain reaction), which detects the virus genome (RNA or DNA), or ELISA (immunoenzymatic assay), which detects virus proteins (Ag—antigens) [4–6].

After a period of infection, antibodies are produced. IgM antibodies are produced first until they reach their maximum in 7 to 10 days. This primary response is taken as an indicator of an acute infection. Subsequently, the secondary immune response occurs faster, more intense, and prolonged. Finally, IgG antibodies will be produced and last longer in the blood. Thus, another approach is detecting the immune response against the virus, the antibodies [5]. It is a method of indirect detection by serological detection because the virus is not detected but the immune response generated by it. These tests are precise and sensitive, usually take a few hours to complete, and require specialized equipment and technical personnel [4–6].

To obtain a faster diagnosis with better specificity and sensitivity, several studies have investigated the use of biosensors from nanomaterials to viral detection [7]. Biosensing devices combine a sensitive biological recognition component (bioreceptor) and a physical transducer. The physical transducer transforms and amplifies the signal from the interaction between the analyte and the bioreceptor into a measurable and quantifiable signal from events such as pH change, heat, light electro, and mass change [3, 4]. Nanomaterials come as an alternative to improve viral detection levels to lower the limit required for detection. Thus, its use can happen in different approaches, such as detecting the virus genome, detecting antigens, and antibodies [7].

Due to the unique properties of the nanomaterials (physic, chemical, electromagnetic, optical, etc.), several classes such as metal noble and carbon-based nanoparticles are being used in biosensors to detect different viral targets, such as enzymes, nucleic acids, antigens, antibodies, and the entire virus. Application of NMs as transducer vary significantly with the quantifiable signal source and primarily have used optical and electrochemical nanobiosensors to virus diagnostic [8, 9].

2.1 Electrochemical Biosensor

Most viral biosensing techniques use nanobiosensors based on electrochemical transductions to produce a useful electrical signal, such as amperometric, potentiometric, voltammetry or impedance sensors [10]. Electrochemical nanosensors contain electrodes, for which the charge distribution, the semiconduction, and dielectric properties are fundamentals. Thus, when the electron production or consumption occurs in a biochemical reaction on the electrode surface, an electrochemical signal is generated [10, 11].

The electrochemical methods display biosensing with selectivity, leading to the ability to measure an individual chemical in the presence of various other chemicals in real-time with high detection accuracy. The general mechanism of sensing is based on the chemisorption of biomarkers that induce a change in conductance due to chemical sensitization [10].

For the detection of different types of viruses, most researchers used electrochemical detection methods, such as amperometry [12], chronoamperometry (CA) [13, 14], conductometry, cyclic voltammetry (CV) [15–20], differential pulse voltammetry (DPV) [9, 15–17, 21], electrochemical impedance spectroscopy (EIS) [15, 20, 22–25] or field effect transistor (FET) [19, 26–28]. Other techniques such as electrochemiluminescence (ECL) [29, 30], linear sweep voltammetry (LSV) [18, 31], and square wave voltammetry (SWV) [19, 32] are also used in viral diagnosis, based on electrochemical transductions by different nanomaterials.

2.2 Optical Biosensor

The photonic devices have wide-range applications in the field of nanobiosensing technology, including in viral detection. Optical biosensors measured the change in the optical characteristics of the transducer surface when the analyte and recognition element form a complex [33]. Different parameters can be used in the detection, such as energy, phase, polarization, absorption, fluorescence, light scattering, amplitude, decay time, and colorimetric [10, 34].

The nanobiosensors based on surface plasmon resonance (SPR) and fluorescence (FL) events are the most common sensors developed and applying for the optical transducer to detection of viruses, including the CoV-2 coronavirus [34–36]. The fluorescence techniques (quenching or enhancing) involve exposure to an external light source to excite the electron transitions in the biomolecules, generating luminescence [34]. Different types of optical biosensors based on light emission have been employed for virus detection, such as chemiluminescence, near-infrared fluorescence (NIR), and luminescence resonance energy transfer (LRET) [34, 37, 38].

The SPR sensor system is based on the light interaction with their surface free electrons that are called "plasmons," which will move far from the atomic core, in a phenomenon called surface plasmon resonance (SPR) [39]. By enhancing or quenching effects of SPR, mainly by localized surface (LSPR), this technique is extremely sensitive, is label-free, and has a rapid real-time detection of high accuracy [34, 40, 41]. Furthermore, NMs based on optical light scattering techniques as dynamic light scattering (DLS) and Raman spectroscopy (surface-enhanced Raman scattering—SERS) have been applied to virus detection [10, 40].

Colorimetric assays involve identifying the target molecules in tested specimens through color changes of an indicator (e.g., metallic NPs and dye molecules). This technique allows a naked eye observation of the presence of biomarkers and measures the absorbance of the colored compounds at a specific wavelength [34].

The colorimetric test offers the advantages of on-site detection of analytes due to its direct reading, convenient operation, and minimum instrumentation requirement [34]. In the present context of a pandemic caused by COVID-19, a simple and fast colorimetric assay is fundamental for a rapid diagnostic and has been utilized through NMs in standard methods such as RT-PCR and LFIA [42–44].

2.3 *Other Biosensors*

In addition to the electrochemical and optical transduction methods mainly used for biosensing, other techniques based on NMs have been adopted to recognize and amplify different signals in viral detection.

Mass change-based biosensing is a promising alternative to viral detection, as by piezoelectric sensors that convert the mechanical energy into an electrical signal, i.e., bulk wave (BW) and surface acoustic wave (SAW) [10]. In this sense, the quartz crystal microbalance (QCM) nanobiosensors measure a mass variation by changing the frequency of a piezoelectric quartz crystal resonator when detecting the biochemical entities [45, 46]. The NMs have been applied to amplification tags to enhance size, elemental, and mass amplification signal transduction through several biosensing techniques [10, 34].

Optomagnetic nanobiosensors, which combine magnetic properties and surface plasmonic enhancement, have been applied in ultra-sensitive biosensing to viral detection. The optomagnetic technique measures the time-dependent voltage by a light-emitting diode in response to an applied oscillating magnetic field from a photodetector illuminated through the magnetic nanoparticles (MNPs) sample [47, 48]. Other rapid and sensitive techniques of viral detection based on magnetism effects are also developed, as magnetic particle spectroscopy (MPS) besides different magnetic immunoassay [49, 50]. Furthermore, MNPs have been used to immobilize and extract biomarkers by a magnetic field, using the standard RT-PCR procedure [50].

3 Nanomaterials in Viral Detection

Nanomaterials (NMs) and nanotechnology allow us to understand the chemistry and physics of different materials by investigating the fundamental substructures that make up matter in an approach known as "bottom-up" science [51]. Due to their nanometric size (10^{-9} m), close to the atomic dimension, the NMs are structurally characterized by having a significant fraction of atoms on the surface. Consequently, it can significantly change their properties compared to conventional coarse-grained polycrystalline materials. These properties strongly depend on the size, morphology, and crystal structure of NMs [52].

NMs include distinct shape as nanoparticles (NPs), nanoclusters (NCs), nanotubes (NTs), nanofibers (NFs), nanowires (NWs), nanorods (NRs), nanofilms, nanocrystals and quantum dots (QDs), with variable composition. Compared to bulk materials, NMs show notable differences in functional properties, such as unique optical, electrical, catalytic, thermal, and magnetic characteristics, due to their minimal size [53]. Such effects have their origin in quantum mechanics, as in the quantum confinement of particles on the surface, giving rise to high chemical reactivity, allowing efficient entrance into the living system [51, 54].

NMs in viral detection comes from their abilities to amplify the signaling events, with suitable surface chemistry to perform the desired molecular bioconjugation, acting as transductors for biosensors [7, 8]. Biosensors are generally classified according to the transduction method they use. The transductor has the role that transforms the signal resulting from the interaction between the analyte and the biological element into a measurable and quantifiable signal [8, 55].

Transductors are often described in their sensitivity to input signals or responsivity and vary significantly with the quantifiable signal source—electricity, magnetism, light, sound, heat, radiation, voltage, vibrations, pressure, and acceleration [56]. In this sense, electrochemical, optical, piezoelectric, magnetic, and thermal nanobiosensors are developed. By fulfilling this role, some NMs stand out as promising candidates, such as carbon-based NMs (CNTs, GO, RGO, GQDs) and metal noble NPs (AuNPs, AgNPs), the applications of which are strongly influenced by their defects and chemical functionalization on their surfaces [3, 8, 54, 56, 57].

Surface modification of the nanostructures or nanomaterials must modify the functional layer to obtain definite selectivity for a given sensing surface. Various studies have been performed to enhance the performance of nanobiosensors by incorporating nanomaterials in the surface to increase the volume-to-surface area ratio and improve the selectivity of the surface [10].

An overview of the main NMs used recently in biosensing devices to detect different viruses is presented. Figure 2 presents a schematic diagram of different nanostructures (NMs) used in biosensing for viral detection, the bioreceptor species, and the main transduction methods reported recently in the literature.

Besides their use in recognition and signal amplification of biosensing, NMs can be applied for the isolation/immobilization and extraction of the bioreceptors in standard detection techniques such as RT-PCR and ELISA. The NMs involved in

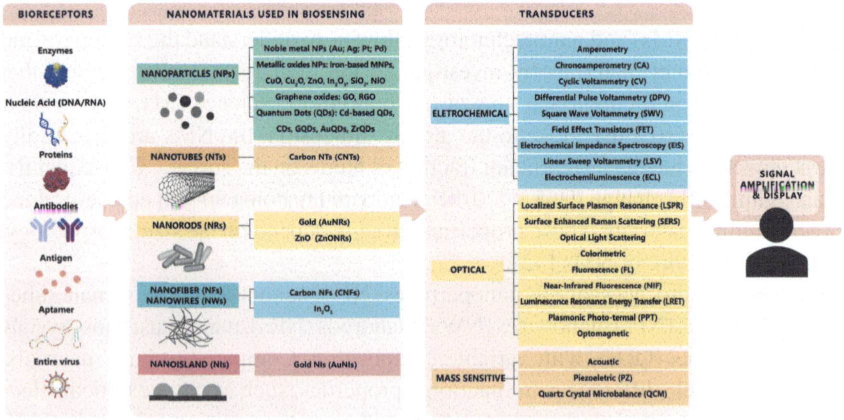

Fig. 2 A scheme of viral biosensor concept based on using the bioreceptors and the transducers, based on different shapes and classes of nanomaterials. Adapted from Alhalaili et al. [34]

the viral analysis include nanoarrays, protein arrays, and nanopore technology but primarily involve nanoparticles [4].

An overview of the recent studies that deal with NMs used in virus detection alternatives is covered in the following topics, focusing on biosensing methods.

3.1 Carbon Nanomaterials in Viral Detection

Carbon nanomaterials in their different allotropes and shapes have been applied to develop new biosensors, improving sensor functionality and recognition [57, 58]. As illustrated in Fig. 3, graphene, graphene oxide (GO), reduced graphene oxide (RGO), carbon nanotubes (CNTs), carbon dots (CDs), and graphene quantum dot (GQD) are

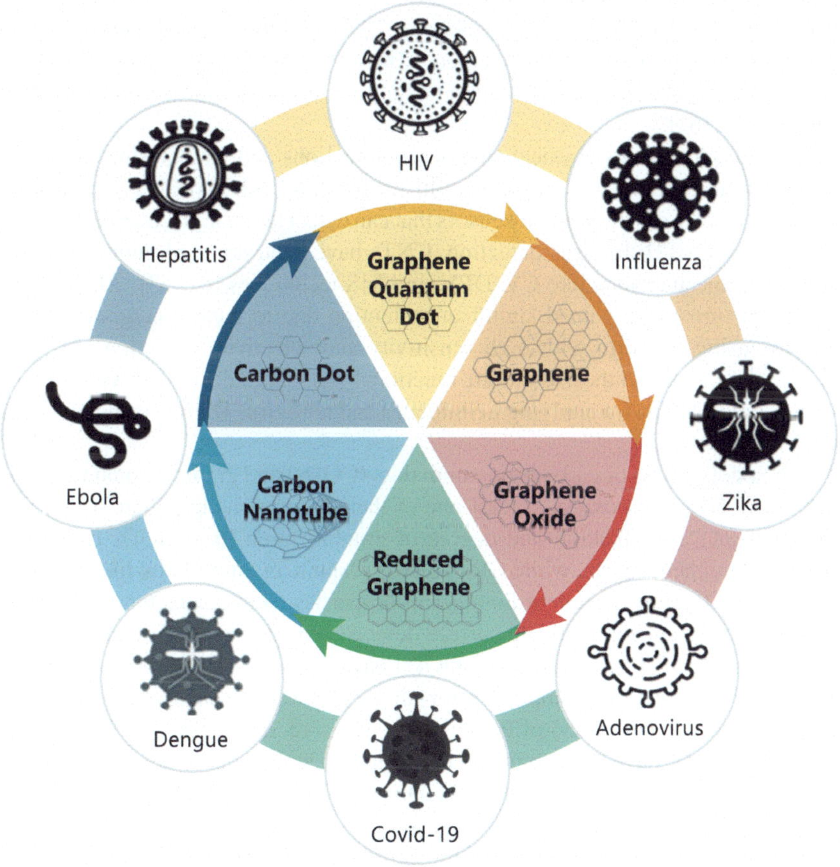

Fig. 3 Carbon-based nanomaterials used for detection of different human viruses. Adapted from Ehtesabi et al. [57]

the most promising nanostructures carbon materials used in applications for the human detection viruses, which each allotrope has its specific functional properties and advantages as a transducer element, besides different biocompatibility and cost conditions [3, 57–59].

The advent of graphene sheets allowed in-depth investigations into their structure and derivates, as the oxides. Graphene oxide (GO) is a single atom carbon layer in which both surfaces have undergone modification with oxygen-bearing functional groups, as carboxylic, phenol hydroxyl, and epoxide groups, being typically manufactured by exfoliation and oxidation of layered crystalline graphite (normal or synthetic), [3, 57]. RGO is obtained from the reduction of GO to graphene through various chemical, thermal or electrochemical paths that maintain some functional groups enabling chemical functionalization to immobilize the molecular receptors onto its surface [57]. Because of this, electrochemical and optical sensing systems bases on GO and RGO have been widely applied in biosensors that capture the interaction (signal) of these NMs with the genetic material viral (DNA or RNA). As a recent example, Wen et al. developed a DNA biosensor based on the fluorescence quenching property of GO to detect EBOV [60], and Zhao et al. developed an ultrasensitive electrochemical detection technology using GO for targeting RNA of SARS-CoV-2 [61].

From transduction of physical/chemical interactions and large surface area, the carbon nanotubes (CNTs), as single or multi-walled (SWCNT/MWCNT), are playing an essential role in assembling biosensors that can detect target molecular biomarkers in trace amounts [3]. In virus detection, CNTs have been reported mainly in electrochemical mechanisms, like CV, DPV, and EIS techniques [62, 63]. Also, CNTs hybrid structures have gained significant attention due to their synergistic effects as a biosensing platform with high sensitivity and selectivity detection [31]. For example, Lee et al. synthesized multi-functionalized CNTs decorated with AuNPs and iron-oxide MNPs for applying in influenza and norovirus DNA sensing channels [31].

Carbon dots (CDs) or carbons quantum dots (CQDs) and graphene quantum dots (GQDs) are NPs that possess at least one of their dimensions less than 10 nm and exhibit distinct hybridization and degree of graphitization [64]. GQDs consist of one or 2–3 graphene layers, while CDs possess graphite or amorphous-like structure [64]. These NPs show excellent fluorescent properties (semiconductor characteristics) due to quantum confinement, surface defects, and edge effects, mainly through doping. [58, 65]. For viral detection, GQDs have been used combined with silver NPs (AuNPs) as the nanocomposites for ultrasensitive and selective electrochemical detection (CV/DPV/EIS) of hepatitis C virus core antigen (HCV) [15]. In the same sense, GQDs have been conjugated with gold nanobundles to detect fowl adenoviruses (FAdVs) by optical methods (FL) through light-matter interaction between both NMs [65].

Other shapes and combinations of the carbon-based NMs are also being developed for viral detection in this field. A carbon nanofiber (CNF) is one of the materials that showed excellent applications in electrochemical biosensors due to its large surface area, stability, and ease of functionalization [19]. When conjugated with metal substrates, the graphene films generally display charge transfer with the metal surface [9]. In addition to these allotropes, diamond nanocrystalline (nanodiamond) have been used in biosensing, such as substrate in electrodes to electrochemical immunoassays and biomolecular techniques [24, 25]. Recently studies developed a rapid-response ultrasensitive biosensor using antibody modified boron-doped nanodiamond for influenza virus detection [24, 25].

The application of carbon-based nanobiosensors holds great promise for the viral detection of more targets/biomarkers. It has great potential in disease diagnostics and clinical analysis, where each allotrope has its specific properties and advantages as a transducer element [59]. Some of the carbon-based NMs applied in biosensors to viral detection reported recently in the literature are summarized in Table 1.

3.2 *Noble Metal Nanomaterials in Viral Detection*

Among the class of NMs, metal noble nanoparticles have been extensively investigated and applied in viral diagnostic, due to their outstanding electrical/optical properties and excellent conditions of chemical stability (low reactivity), even under conditions of high surface area such as in nanostructures [59] (Table 2). In this sense, nanoparticles of gold (AuNPs) and silver (AgNPs) are the main NMs used in virus detection through electrochemical and optical transductions in biosensors.

Gold NPs are most used in biosensing viral detection main due to their catalytic properties, excellent biocompatibility, and simple production pathway [59].

Figure 4 shows a scheme for the basic characteristics and functions of AuNPs applied in virus detection. AuNPs exhibit resonant light scattering, fluorescence quenching or enhancing, and color amplification, allowing numerous optical transductors to develop [10, 39, 42, 70]. Colorimetric virus detection generally uses AuNPs as markers for direct staining (intense red color) or in color change events (red to purple) when particle aggregation [10]. At a recent time, Moitra et al. developed a colorimetric bioassay to SARS-CoV-2 diagnostic, utilizing the anisotropic plasmonic properties of AuNPs for efficient targeting [42]. For the same purpose, Mahari et al. utilized AuNPs through electrochemical mechanisms (CV and DPV), measuring the change in the electrical conductivity for rapid detection of SARS-CoV-2 [17]. Due to their electrical and redox properties, AuNPs are broadly described as electroactive and catalytic tags in various electrochemical techniques applied to virus detection (Fig. 4) [10].

AuNPs have also been utilizing as tags to amplify the signals detection of mass (QCM), size (AFM), and elemental (ICP-MS). The large surface area-to-volume ratio of the AuNPs allows them to act as a scaffold for bioimmobilization and target-probe interactions with specificity and sensitivity (Fig. 4) [10, 46].

Analogously, silver nanoparticles (AgNPs) are promising in creating novel biosensing devices due to their particularities, such as simplicity in producing a color response with high sensitivity [71]. Teengam et al. developed a colorimetric biosensor of low-cost and rapid viral detection, for MERS-CoV and HPV, from

Table 1 Carbon-based nanomaterials used in techniques for viral detection

NMs	Detection method	Disease (Virus)—Target	Refs.
GQD—AgNPs	Electrochemical—CV/DPV/EIS	Hepatitis C (HCV)—Antigen	Valipour and Roushani [15]
GQD—AuNBs (nanobundles)	Optical—FL	Adenoviruses (FAdVs)—Entire virus	Ahmed et al. [65]
GQD	Electrochemical—CV/DPV	Hepatitis B (HBV)—DNA	Xiang et al. [16]
CDs—GO/AuNPs	Optical—FL	AIDS (HIV)—DNA	Qaddare and Salimi [66]
Graphene-Au nanocomposite	Electrochemical—EIS	Influenza A—Entire virus	Anik et al. [22]
Graphene	Electrochemical—FET	Influenza A—Lectin	Ono et al. [26]
Graphene	Electrochemical—FET	COVID-19 (SARS-CoV-2)—Antigen	Seo et al. [27]
Graphene—AuNPs/MNPs	Optical—FL/Optomagnetic	Influenza A (H1N1)—Entire virus	Lee et al. [67]
GO	Electrochemical—EIS	Dengue (DENV)—RNA and DNA	Jin et al. [68]
GO	Electrochemical—DPV	COVID-19 (SARS-CoV-2)—RNA	Zhao et al. [61]
GO	Optical—FL	Ebola (EBOV)—DNA	Wen et al. [60]
RGO	Electrochemical—CA	Influenza A (H1N1)—Entire virus	Singh et al. [14]
RGO	Electrochemical—FET	Influenza A (H5N1)—DNA	Chan et al. [69]
RGO	Electrochemical—EIS	AIDS (HIV-1)—DNA	Shamsipur et al. [23]
MWCNTs	Electrochemical—CV/DPV	AIDS (HIV-1)—Antigen	Ma et al. [62]
CNT	Electrochemical—EIS	Dengue (DENV)—Antibody	Palomar et al. [63]
CNT—AuNPs/MNPs/CdQDs	Electrochemical—LSV	Influenza A (H1N1)—DNA	Lee et al. [31]
CNFs	Electrochemical—CV/SWV/FET	MERS (CoV)—Antigen	Eissa and Zourob [19]
Nanodiamond (boron-doped)	Electrochemical—EIS	Influenza A—Protein	Nidzworski et al. [24]
Nanodiamond (boron-doped)	Electrochemical—EIS	Influenza A (H1N1 and H3N2)—Protein	Siuzdak et al. [25]

Table 2 Noble metal nanomaterials used in techniques for viral detection

NMs	Detection method	Disease (Virus)—Target	Refs.
AgNPs	Optical—Colorimetric	MERS and HPV—DNA	Teengam et al. [72]
AgNCs	Optical—FL	AIDS (HIV-1/HIV-2)—DNA	Zou et al. [76]
AgNPs	Optical—FL	AIDS (HIV-1)—Antigen	Kurdekar et al. [77]
AgNPs + graphene	Electrochemical—CV/LSV	Influenza A (H7)—Entire virus	Huang et al. [18]
AuNPs	Optical—Colorimetric/ LSPR	Influenza A (H5N1)—Antigen	Xu et al. [78]
AuNPs	Electrochemical—CA	Influenza A (H9N2)—Entire virus	Sayhi et al. [13]
AuNPs + RGO	Electrochemical—EIS	AIDS (HIV-1)—DNA	Shamsipur et al. [23]
AuNRs	Optical—FL	Hepatitis B (HBV)—DNA	Negahdari et al. [79]
AuNPs + RGO	Electrochemical—DPV	Hepatitis B (HBV)—DNA	Oliveira et al. [9]
AuNPs	Electrochemical—SWV	MERS (CoV)—Antigen	Layqah and Eissa [32]
AuNPs	Optical—LSPR/Colorimetric/DLS	MERS (CoV)—DNA	Kim et al. [40]
AuNIs	Optical—PPT/LSPR	COVID-19 (SARS-CoV-2)—DNA	Qiu et al. [36]
AuNPs	RT-PCR—Colorimetric	COVID-19 (SARS-CoV-2)—RNA	Kumar et al. [43]
AuNPs	Eletrochemical—CV/DPV	COVID-19 (SARS-CoV-2)—Antigen	Mahari et al. [17]
AuNPs	Optical—Colorimetric/SPR	COVID-19 (SARS-CoV-2)—RNA	Moitra et al. [42]
AuNPs (colloids)	LFIA—Colorimetric	COVID-19 (SARS-CoV-2)—Antibody	Li et al. [80]
AuPtNPs + GO	Electrochemical—DPV	AIDS (HIV-1)—DNA	Akbari et al. [73]
AuPtNPs	Electrochemical—DPV/CV/EIS	COVID-19 (SARS-CoV-2)—Aptamers and Proteins	Tian et al. [74]
PtNPs	Electrochemical/Optical	Hepatitis B (HBV), Hepatitis C (HCV), Zika (ZIKV)—Entire virus	Draz et al. [75]

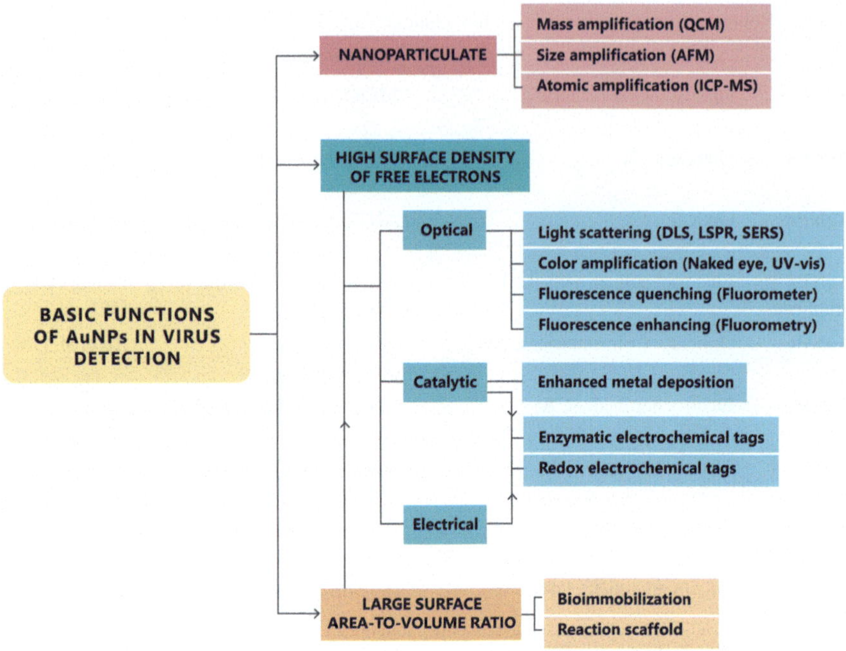

Fig. 4 A generalized scheme for the essential characteristics and functions of AuNPs applied in virus detection. Adapted from Draz et al. [10]. *QCM—quartz crystal microbalance, AFM—atomic force microscopy, ICP-MS—inductively coupled plasma mass spectrometry, DLS—dynamic light scattering, LSPR—localized surface plasmon resonance, SERS—surface-enhanced Raman scattering, UV-vis-ultraviolet-visible*

an interaction mechanism between AgNPs and distinct bioreceptors [72]. Huang et al. designed an electrochemical immunosensor based on AgNPs-coated graphene, bioconjugates for the high sensitivity detection of influenza virus (AIV H7) [18].

Furthermore, the noble metal Platinum (PtNPs) and composites, as with graphene and its oxides, have been evaluated in viral detection [73, 74]. Tian et al. constructed an electrochemical biosensor with high sensitivity and selectivity to detect SARS-CoV-2 through Au-Pt bimetallic nanoparticles [74]. Draz et al. used of PtNPs through a nanoparticle-enabled smartphone (NES) system for rapid, simple, and sensitive virus detection, such as Zika virus (ZIKV), hepatitis B virus (HBV), and HCV. The detection process using a convolutional neural network (CNN)—enabled smartphone system and comprises the capture of target virus and labeling with PtNPs (nanoprobes) that, in contact with hydrogen peroxide (H_2O_2) complex bubbles of oxygen which are detected by screenshots via an optical mechanism (CNN-NES) [75].

3.3 Metallic Oxides Nanomaterials in Viral Detection

Magnetic nanoparticles (MNPs) with proper surface functionalization have been evaluated to labels for magnetic immunoassays as an optomagnetic sensing strategy through the assembly of MNPs clusters with the biological target [49] (Table 3). The optomagnetic sensor detects the presence and increases in the size of MNPs clusters when their dimensions are similar to the wavelength of light through optical transmission measurements in an oscillating magnetic field [47]. Tian et al. produced different works with this detection technique, as in the diagnosis of dengue and covid19 viruses [47, 48]. Via electrochemical methods (CV), Zhang et al. used MNPs as self-sacrificial labels with superior sensitivity and simplicity for accurate sample detection of virus H5N1[81]. In another way, Zhao et al. employed MNPs to viral RNA extraction method for the sensitive detection of COVID-19 causing virus, reduce the turn-around time and operational requirements in current molecular diagnosis (RT-PCR) [50].

Nanostructured zinc oxide (ZnO), an important semiconductor material, has been applied in immunosensors and catalysts for virus detection. ZnO nanomaterial exhibits a large surface area with high porosities and is accessible to modification, which allows supporting other components to form hybrid nanostructures [21]. For example, Yang et al. evaluated a nanohybrid ZnO-based immunosensor to detect influenza through the surface functionalization of porous ZnO spheres (pZnO) with

Table 3 Metallic oxides nanomaterials used in techniques for viral detection

NMs	Detection method	Disease (Virus)—Target	Refs.
MNPs	MPS—magnetic particle spectroscopy	Infuenza A (H1N1)—Antigen	Wu et al. [49]
MNPs	Optomagnetic	Dengue (DENV)—DNA	Tian et al. [48]
MNPs	Optomagnetic	COVID-19 (SARS-CoV-2)—DNA	Tian et al. [47]
MNPs (Fe_3O_4)	Electrochemical—CV	Influenza A (H5N1)—Entire virus	Zhang et al. [81]
MNPs	Magnetic immunoassay—(RT-PCR)	COVID-19 (SARS-CoV-2)—RNA	Zhao [50]
NiO thin films	Electrochemical—FET	AIDS (HIV-1)—DNA	Mansouri et al. [28]
ZnO NPs—Pt(Au/Ag)	Electrochemical—DPV	Influenza—Entire virus	Yang et al. [21]
ZnO NRs	Electrochemical—Amperometry	Influenza A (H1N1, H5N1 and H7N9)—Antigen	Han et al. [12]
Cu_2O NPs (nanocubes)	Electrochemical—CV/EIS	COVID-19 (SARS-CoV-2)—Protein	Rahmati et al. [20]

PtNPs, comparing the catalytical property with AuNPs and AgNPs [21]. Furthermore, from another shape, ZnO nanorods (ZnONRs) were designed in an electrochemical immunosensor chip for simultaneous multiple antigen/biomarkers sensing, including influenza A viruses [12].

Other metallic nanomaterials have been evaluated for viral detection, such as copper and nickel oxides, from different shapes and compositions [20, 28]. Through electrochemical nanodevices, Cu_2O nanocubes have been employed to increase the biomarkers loading on a carbon electrodes surface to detect the SARS-CoV-2 virus (Covid-19) in less than 20 min to facilitating the point-of-care diagnosis [20]. Furthermore, to improve the selectivity toward HIV gene detection, a p-type semiconducting nickel oxide (NiO) nanofilm has been developed in the FET type-biosensor [28].

3.4 Other Nanomaterials in Viral Detection

Metallic and carbon-based nanomaterials are the most used in viral detection techniques. Moreover, other nanomaterials have been widely developed and evaluated aimed at optimizing the diagnostic of distinct viruses (Table 4).

Selenium nanoparticles (SeNPs) have been employed in an immunoassay (LFIA), which detects the Cov-2 IgM and IgG antibodies in a short time (10 min) [82]. While qualitative probes for LFIA procedure, SeNPs have exceptional levels of stability and sensitivity and are more economical to produce than colloidal gold [82]. Also, selenium has been assembled through quantum dots and nanocomposites [41]. In this sense, Nasrin et al. developed a biosensor by fluorescence QDs (CdZnSeS/ZnSeS) and AuNPs, using tunable LSPR-based fluorometric techniques [41].

Different compositions of quantum dots (QDs) and their composites have been applied to viral detection. Cadmium (Cd)-based QDs, in addition to other elements such as tellurium (Te), zinc (Zn), and sulfur (S), make up a new class of nanomaterials that has been widely investigated for the detection of several viruses [30, 38, 41]. For example, a ratiometric fluorescence biosensor utilizing CdTe-carbon dots (CdTeCDs) and quantum dots (CdTeQDs) have been evaluated for a rapid and specific viral detection [83].

Nanoparticles consisting of rare earth elements have also been synthesized and evaluated for virus detection. For example, recent studies developed a novel electrochemiluminescence biosensor for the highly sensitive and selective HIV gene detection using Europium sulfide nanocrystals (EuS NCs) as a signal-producing compound. In this sense, Chen et al. developed a rapid and sensitive method to detect Cov-2 IgG antibodies, using Lanthanide (La)-doped NPs by LFIA procedure [44].

Thus, nanomaterials in varieties of compositions and shapes are presented as a promising alternative in components for detecting different types of viruses. NMs are widely used in standard tests such as RT-PCR and immunoassays (ELISA and LFIA) in bio-immobilization and bio-probe processes. However, in the biosensing

Table 4 Others nanomaterials used in techniques for viral detection

NMs	Detection method	Disease (Virus)—Target	Refs.
CdTe-CDs	Optical—FL	AIDS (HIV)—DNA	Liang et al. [83]
CdTe-QDs	Optical—FL	AIDS (HIV)—DNA	Deng et al. [84]
CdZnSeTeS QDs	Optical—NIR/FL	Influenza A (H1N1)—RNA	Adegoke et al. [38]
CdTe QDs - AuNPs	Electrochemical—ECL	Hepatitis B (HBV) and Hepatitis C (HCV)—DNA	Liu et al. [30]
CdSe@ZnS QDs	LFS (lateral flow strips)—FL	COVID-19 (SARS-CoV-2)—Antibody Influenza A (H5N1)—Entire virus	Li et al. [35]
CdZnSeS/ZnSeS QDs—AuNPs	Optical—LSPR/FL	Influenza A (H1N1)—Antigen	Nasrin et al. [41]
ZrQDs/ZrNPs	Optical—FL	Infectious bronchitis (IBV)—Entire virus	Ahmed et al. [85]
SeNPs	LFIA—Colorimetric/RT-PCR	COVID-19 (SARS-CoV-2)—Antibody	Wang et al. [82]
EuS nanocrystals	Eletrochemical—ECL	AIDS (HIV-1)—DNA	Babamiri et al. [29]
Lanthanide-doped polystyrene nanoparticles (LNPs)	LFIA—Colorimetric	COVID-19 (SARS-CoV-2)—Antibody	Chen et al. [44]

field, NMs demonstrate a wide application possibility from developing transduction elements, mainly electrochemical and optical.

4 Conclusion

In the constant search for a sensitive, accurate, and primarily rapid virus diagnostic, the nanomaterials emerge as a potential solution, acting mainly as transducers in different biosensing techniques to recognize and amplify signals. The application of NMs is strongly influenced by their defects and chemical functionalization on their large surface area. The detection by biosensing techniques predominantly involves electrochemical and optical sensors developed through several types of nanomaterials.

In this sense, the main NMs used are the gold nanoparticles AuNPs, and other noble metal (Ag, Pt) NPs, due to their plasmon surface that enable distinct transduction strategies such as fluorescence, light scattering, and colorimetric, besides the enhance of the electrical signal in electronic circuits. The use of carbon-based

NMs for biosensors is also widespread, such as graphene and its compounds (GO, RGO and GQDs) and carbon nanotubes (CNTs). Metal oxide nanoparticles such as ZnONPs and iron-based MNPs are also used, including in steps of extraction and immobilization of biomarkers. A significant increase in the development and application of several combinations of nanocomposites was observed, mainly between AuNPs and carbon-based NMs. Also, quantum dots (QDs) are based on cadmium (Cd), tellurium (Te), zinc (Zn), and zirconium (Zr), in addition to nanoparticles of selenium (SeNPs) and rare earth, have been studied for viral detection. It is expected that nanomaterials can offer practical and viable solutions in enhancing the sensitivity and selectivity of viral detection devices, allowing diagnosis in a fast and efficient procedure.

References

1. Balloux, F., van Dorp, L.: Q&A: What are pathogens, and what have they done to and for us? BMC Biol. **15**, 4–9 (2017). https://doi.org/10.1186/s12915-017-0433-z
2. Nasrollahzadeh, M., Sajjadi, M., Soufi, G.J., Iravani, S., Varma, R.S.: Nanomaterials and nanotechnology-associated innovations against viral infections focus on coronaviruses. Nanomaterials **10** (2020). https://doi.org/10.3390/nano10061072
3. Mokhtarzadeh, A., Eivazzadeh-Keihan, R., Pashazadeh, P., Hejazi, M., Gharaatifar, N., Hasanzadeh, M., Baradaran, B., de la Guardia, M.: Nanomaterial-based biosensors for detection of a pathogenic virus. TrAC—Trends Anal. Chem. **97**, 445–457 (2017). https://doi.org/10.1016/j.trac.2017.10.005
4. Vaculovicova, M., Michalek, P., Krizkova, S., Macka, M., Adam, V.: Nanotechnology-based analytical approaches for detection of viruses. Anal. Methods. **9**, 2375–2391 (2017). https://doi.org/10.1039/c7ay00048k
5. Brazaca, L.C., dos Santos, P.L., de Oliveira, P.R., Rocha, D.P., Stefano, J.S., Kalinke, C., Abarza Muñoz, R.A., Bonacin, J.A., Janegitz, B.C., Carrilho, E.: Biosensing strategies for the electrochemical detection of viruses and viral diseases—A review. Anal. Chim. Acta. **1159** (2021). https://doi.org/10.1016/j.aca.2021.338384
6. Bustin, S.A., Mueller, R.: Real-time reverse transcription PCR (qRT-PCR) and its potential use in clinical diagnosis. Clin. Sci. **109**, 365–379 (2005). https://doi.org/10.1042/CS20050086
7. Hassanzadeh, P.: Nanotheranostics against COVID-19: From multivalent to immune-targeted materials. J. Control. Release. **328**, 112–126 (2020). https://doi.org/10.1016/j.jconrel.2020.08.060
8. Palestino, G., García-Silva, I., González-Ortega, O., Rosales-Mendoza, S.: Can nanotechnology help in the fight against COVID-19? Expert Rev. Anti. Infect. Ther. **18**, 849–864 (2020). https://doi.org/10.1080/14787210.2020.1776115
9. Oliveira, D.A., Silva, J.V., Flauzino, J.M.R., Castro, A.C.H., Moço, A.C.R., Soares, M.M.C.N., Madurro, J.M., Brito-Madurro, A.G.: Application of nanomaterials for the electrical and optical detection of the hepatitis B virus. Anal. Biochem. **549**, 157–163 (2018). https://doi.org/10.1016/j.ab.2018.03.023
10. Draz, M.S., Shafiee, H.: Applications of gold nanoparticles in virus detection. Theranostics. **8**, 1985–2017 (2018). https://doi.org/10.7150/thno.23856
11. Malik, P., Katyal, V., Malik, V., Asatkar, A., Inwati, G., Mukherjee, T.K.: Nanobiosensors: Concepts and Variations. ISRN Nanomater. **2013**, 1–9 (2013). https://doi.org/10.1155/2013/327435
12. Han, J.H., Lee, D., Chew, C.H.C., Kim, T., Pak, J.J.: A multi-virus detectable microfluidic electrochemical immunosensor for simultaneous detection of H1N1, H5N1, and H7N9 virus

using ZnO nanorods for sensitivity enhancement. Sensors Actuators, B Chem. **228**, 36–42 (2016). https://doi.org/10.1016/j.snb.2015.07.068
13. Sayhi, M., Ouerghi, O., Belgacem, K., Arbi, M., Tepeli, Y.: Biosensors and Bioelectronics Electrochemical detection of influenza virus H9N2 based on both immunomagnetic extraction and gold catalysis using an immobilization-free screen-printed carbon microelectrode. Biosens. Bioelectron. **107**, 170–177 (2018). https://doi.org/10.1016/j.bios.2018.02.018
14. Singh, R., Hong, S., Jang, J.: Label-free Detection of Influenza Viruses using a Reduced Graphene Oxide-based Electrochemical Immunosensor Integrated with a Microfluidic Platform. Sci. Rep. **7**, 1–11 (2017). https://doi.org/10.1038/srep42771
15. Valipour, A., Roushani, M.: Using silver nanoparticle and thiol graphene quantum dots nanocomposite as a substratum to load antibody for detection of hepatitis C virus core antigen: Electrochemical oxidation of riboflavin was used as redox probe. Biosens. Bioelectron. **89**, 946–951 (2017). https://doi.org/10.1016/j.bios.2016.09.086
16. Xiang, Q., Huang, J., Huang, H., Mao, W., Ye, Z.: A label-free electrochemical platform for the highly sensitive detection of hepatitis B virus DNA using graphene quantum dots. RSC Adv. **8**, 1820–1825 (2018). https://doi.org/10.1039/c7ra11945c
17. Mahari, S., Roberts, A., Shahdeo, D., Gandhi, S.: eCovSens-Ultrasensitive Novel In-House Built Printed Circuit Board Based Electrochemical Device for Rapid Detection of nCovid-19 antigen, a spike protein domain 1 of SARS-CoV-2 (2020)
18. Huang, J., Xie, Z., Xie, Z., Luo, S., Xie, L., Huang, L., Fan, Q., Zhang, Y., Wang, S., Zeng, T.: Silver nanoparticles coated graphene electrochemical sensor for the ultrasensitive analysis of avian influenza virus H7. Anal. Chim. Acta. **913**, 121–127 (2016). https://doi.org/10.1016/j.aca.2016.01.050
19. Eissa, S., Zourob, M.: Development of a low-cost cotton-tipped electrochemical immunosensor for the detection of SARS-CoV-2. Anal. Chem. **93**, 1826–1833 (2021). https://doi.org/10.1021/acs.analchem.0c04719
20. Rahmati, Z., Roushani, M., Hosseini, H., Choobin, H.: Electrochemical immunosensor with Cu2O nanocube coating for detection of SARS-CoV-2 spike protein. Microchim. Acta. **188** (2021). https://doi.org/10.1007/s00604-021-04762-9
21. Yang, Z., Zhuo, Y., Yuan, R., Chai, Y.: Biosensors and Bioelectronics A nanohybrid of platinum nanoparticles-porous ZnO—hemin with electrocatalytic activity to construct an amplified immunosensor for detection of influenza **78**, 321–327 (2016). https://doi.org/10.1016/j.bios.2015.10.073
22. Anik, U., Tepeli, Y., Sayhi, M., Nsiri, J., Diouani, M.F.: Towards the electrochemical diagnostic of influenza virus: Development of a graphene-Au hybrid nanocomposite modified influenza virus biosensor based on neuraminidase activity. Analyst. **143**, 150–156 (2018). https://doi.org/10.1039/c7an01537b
23. Shamsipur, M., Samandari, L., Taherpour, A.A., Pashabadi, A.: Analytica Chimica Acta Subfemtomolar detection of HIV-1 gene using DNA immobilized on composite platform reinforced by a conductive polymer sandwiched between two nanostructured layers : A solid signal-amplification strategy. Anal. Chim. Acta. **1055**, 7–16 (2019). https://doi.org/10.1016/j.aca.2018.12.013
24. Nidzworski, D., Siuzdak, K., Niedziałkowski, P., Bogdanowicz, R., Sobaszek, M., Ryl, J., Weiher, P., Sawczak, M., Wnuk, E., Goddard, W.A., Jaramillo-Botero, A., Ossowski, T.: A rapid-response ultrasensitive biosensor for influenza virus detection using antibody modified boron-doped diamond. Sci. Rep. **7**, 1–10 (2017). https://doi.org/10.1038/s41598-017-15806-7
25. Siuzdak, K., Niedziałkowski, P., Sobaszek, M., Łęga, T., Sawczak, M., Czaczyk, E., Dziąbowska, K., Ossowski, T., Nidzworski, D., Bogdanowicz, R.: Biomolecular influenza virus detection based on the electrochemical impedance spectroscopy using the nanocrystalline boron-doped diamond electrodes with covalently bound antibodies. Sens. Actuators, B Chem. **280**, 263–271 (2019). https://doi.org/10.1016/j.snb.2018.10.005
26. Ono, T., Oe, T., Kanai, Y., Ikuta, T., Ohno, Y., Maehashi, K., Inoue, K., Watanabe, Y., Nakakita, S.I., Suzuki, Y., Kawahara, T., Matsumoto, K.: Glycan-functionalized graphene-FETs toward selective detection of human-infectious avian influenza virus. Jpn. J. Appl. Phys. **56** (2017). https://doi.org/10.7567/JJAP.56.030302

27. Seo, G., Lee, G., Kim, M.J., Baek, S.H., Choi, M., Ku, K.B., Lee, C.S., Jun, S., Park, D., Kim, H.G., Kim, S.J., Lee, J.O., Kim, B.T., Park, E.C., Il Kim, S.: Rapid detection of COVID-19 causative virus (SARS-CoV-2) in human nasopharyngeal swab specimens using field-effect transistor-based biosensor. ACS Nano. **14**, 5135–5142 (2020). https://doi.org/10.1021/acsnano.0c02823.
28. Mansouri, S., Salimi, A., Astinchap, B.: Sensors and Actuators B : Chemical The development of radio frequency magnetron sputtered p-type nickel oxide thin film field-effect transistor device combined with nucleic acid probe for ultrasensitive label-free HIV-1 gene detection. Sens. Actuators B. Chem. **266**, 178–186 (2018). https://doi.org/10.1016/j.snb.2018.03.111
29. Babamiri, B., Salimi, A., Hallaj, R.: Biosensors and Bioelectronics A molecularly imprinted electrochemiluminescence sensor for ultrasensitive HIV-1 gene detection using EuS nanocrystals as luminophore. Biosens. Bioelectron. **117**, 332–339 (2018). https://doi.org/10.1016/j.bios.2018.06.003
30. Liu, L., Wang, X., Ma, Q., Lin, Z., Chen, S., Li, Y., Lu, L., Qu, H., Su, X.: Multiplex electrochemiluminescence DNA sensor for determination of hepatitis B virus and hepatitis C virus based on multicolor quantum dots and Au nanoparticles. Anal. Chim. Acta. **916**, 92–101 (2016). https://doi.org/10.1016/j.aca.2016.02.024
31. Lee, J., Morita, M., Takemura, K., Park, E.Y.: A multi-functional gold/iron-oxide nanoparticle-CNT hybrid nanomaterial as virus DNA sensing platform. Biosens. Bioelectron. **102**, 425–431 (2018). https://doi.org/10.1016/j.bios.2017.11.052
32. Layqah, L.A., Eissa, S.: An electrochemical immunosensor for the coronavirus associated with the Middle East respiratory syndrome using an array of gold nanoparticle-modified carbon electrodes. Microchim. Acta. **186**, 2–11 (2019). https://doi.org/10.1007/s00604-019-3345-5
33. Saylan, Y., Erdem, Ö ., Ünal, S., Denizli, A.: An alternative medical diagnosis method: Biosensors for virus detection. Biosensors **9** (2019). https://doi.org/10.3390/bios9020065.
34. Alhalaili, B., Popescu, I.N., Kamoun, O., Alzubi, F.: Coronavirus 2019-nCoV and Other. Sensors **20**, 1–45 (2020)
35. Li, C., Zou, Z., Liu, H., Jin, Y., Li, G., Yuan, C., Xiao, Z., Jin, M.: Synthesis of polystyrene-based fluorescent quantum dots nanolabel and its performance in H5N1 virus and SARS-CoV-2 antibody sensing. Talanta **225**, 122064 (2021). https://doi.org/10.1016/j.talanta.2020.122064
36. Qiu, G., Gai, Z., Tao, Y., Schmitt, J., Kullak-Ublick, G.A., Wang, J.: Dual-Functional Plasmonic Photothermal Biosensors for Highly Accurate Severe Acute Respiratory Syndrome Coronavirus 2 Detection. ACS Nano **14**, 5268–5277 (2020). https://doi.org/10.1021/acsnano.0c02439
37. Ye, W.W., Tsang, M.K., Liu, X., Yang, M., Hao, J.: Upconversion luminescence resonance energy transfer (LRET)-based biosensor for rapid and ultrasensitive detection of avian influenza virus H7 subtype. Small. **10**, 2390–2397 (2014). https://doi.org/10.1002/smll.201303766
38. Adegoke, O., Kato, T., Park, E.Y.: An ultrasensitive alloyed near-infrared quinternary quantum dot-molecular beacon nanodiagnostic bioprobe for influenza virus RNA. Biosens. Bioelectron. **80**, 483–490 (2016). https://doi.org/10.1016/j.bios.2016.02.020
39. Rabiee, N., Bagherzadeh, M., Ghasemi, A., Zare, H.: Point-of-Use Rapid Detection of SARS-CoV-2 : Nanotechnology-Enabled Solutions for the COVID-19 Pandemic (2020)
40. Kim, H., Park, M., Hwang, J., Kim, J.H., Chung, D.: Development of Label-Free Colorimetric Assay for MERS-CoV Using (2019). https://doi.org/10.1021/acssensors.9b00175
41. Nasrin, F., Chowdhury, A.D., Takemura, K., Kozaki, I., Honda, H., Adegoke, O., Park, E.Y.: Fluorometric virus detection platform using quantum dots-gold nanocomposites optimizing the linker length variation. Anal. Chim. Acta. **1109**, 148–157 (2020). https://doi.org/10.1016/j.aca.2020.02.039
42. Moitra, P., Alafeef, M., Dighe, K., Frieman, M.B., Pan, D.: Selective Naked-Eye Detection of SARS-CoV - 2 Mediated by N Gene Targeted Antisense Oligonucleotide Capped Plasmonic Nanoparticles (2020) doi:https://doi.org/10.1021/acsnano.0c03822.
43. Kumar, V., Mishra, S., Sharma, R., Agarwal, J., Ghoshal, U., Khanna, T., Sharma, L.K., Verma, S.K., Mishra, P., Medicine, M., Pgims, S.G., Road, R., Sciences, M., Pgims, S.G., Road, R., Council, I., Bhawan, R., Delhi, N., Informatics, H., Pgims, S.G., Road, R., Tiwari, S., Medicine,

M., Science, M.: Development of RNA-based assay for rapid detection of SARS-CoV-2 in clinical samples (2020)

44. Chen, Z., Zhang, Z., Zhai, X., Li, Y., Lin, L., Zhao, H., Bian, L., Li, P., Yu, L., Wu, Y., Lin, G.: Rapid and Sensitive Detection of anti-SARS-CoV-2 IgG, Using Lanthanide-Doped Nanoparticles-Based Lateral Flow Immunoassay. Anal. Chem. **92**, 7226–7231 (2020). https://doi.org/10.1021/acs.analchem.0c00784
45. Pandey, L.M.: Design of engineered surfaces for prospective detection of Sars-CoV-2 using quartz crystal microbalance based techniques. Expert Rev. Proteomics. **17**, 425–432 (2020). https://doi.org/10.1080/14789450.2020.1794831
46. Yuan, Y.J., Liang, T., Han, K.: Development of a real-time QCM bond-rupture system for POCT applications. IEEE Sens. J. **16**, 8731–8735 (2016). https://doi.org/10.1109/JSEN.2016.2563167
47. Tian, B., Gao, F., Fock, J., Dufva, M., Fougt, M.: Biosensors and Bioelectronics Homogeneous circle-to-circle amplification for real-time optomagnetic detection of SARS-CoV-2 RdRp coding sequence. Biosens. Bioelectron. **165**, 112356 (2020). https://doi.org/10.1016/j.bios.2020.112356
48. Tian, B., Fock, J., Antonio, G., Minero, S., Fougt, M.: Biosensors and Bioelectronics Nicking-assisted on-loop and off-loop enzymatic cascade amplification for optomagnetic detection of a highly conserved dengue virus sequence. Biosens. Bioelectron. **160**, 112219 (2020). https://doi.org/10.1016/j.bios.2020.112219
49. Wu, K., Liu, J., Saha, R., Su, D., Krishna, V.D., Cheeran, M.C.J., Wang, J.P.: Magnetic Particle Spectroscopy for Detection of Influenza A Virus Subtype H1N1. ACS Appl. Mater. Interfaces. **12**, 13686–13697 (2020). https://doi.org/10.1021/acsami.0c00815
50. Zhao, Z.: A simple magnetic nanoparticles-based viral RNA extraction method for efficient detection of SARS-CoV-2, p. 518055 (2020)
51. William, J., Callister, D.: Materials Science and Engineering, 7th edn. An Introduction (2007). https://doi.org/10.1007/BF01184995
52. Meyers, M.A., Mishra, A., Benson, D.J.: Mechanical properties of nanocrystalline materials. Prog. Mater. Sci. **51**, 427–556 (2006). https://doi.org/10.1016/j.pmatsci.2005.08.003
53. Farré, M., Sanchís, J., Barceló, D.: Analysis and assessment of the occurrence, the fate and the behavior of nanomaterials in the environment. TrAC - Trends Anal. Chem. (2011). https://doi.org/10.1016/j.trac.2010.11.014
54. Kang, J., Tahir, A., Wang, H., Chang, J.: Applications of nanotechnology in virus detection, tracking, and infection mechanisms. Wiley Interdiscip. Rev. Nanomed. Nanobiotechnol. (2021). https://doi.org/10.1002/wnan.1700
55. Rocchitta, G., Spanu, A., Babudieri, S., Latte, G., Madeddu, G., Galleri, G., Nuvoli, S., Bagella, P., Demartis, M.I., Fiore, V., Manetti, R., Serra, P.A.: Enzyme biosensors for biomedical applications: Strategies for safeguarding analytical performances in biological fluids. Sensors (Switzerland) **16** (2016). https://doi.org/10.3390/s16060780.
56. Lowe, C.R.: Overview of Biosensor and Bioarray Technologies. Handb. Biosens. Biochips. (2008). https://doi.org/10.1002/9780470061565.hbb003
57. Ehtesabi, H.: Application of carbon nanomaterials in human virus detection. J. Sci. Adv. Mater. Devices. **5**, 436–450 (2020). https://doi.org/10.1016/j.jsamd.2020.09.005
58. Farzin, L., Shamsipur, M., Samandari, L., Sheibani, S.: HIV biosensors for early diagnosis of infection: The intertwine of nanotechnology with sensing strategies. Talanta **206**, 120201 (2020). https://doi.org/10.1016/j.talanta.2019.120201
59. Antiochia, R.: Nanobiosensors as new diagnostic tools for SARS, MERS and COVID-19: from past to perspectives. Microchim. Acta. **187** (2020). https://doi.org/10.1007/s00604-020-04615-x.
60. Wen, J., Li, W., Li, J., Tao, B., Xu, Y., Li, H., Lu, A., Sun, S.: Sensors and Actuators B : Chemical Study on rolling circle amplification of Ebola virus and fluorescence detection based on graphene oxide. Sensors Actuators B. Chem. **227**, 655–659 (2016). https://doi.org/10.1016/j.snb.2016.01.036

61. Zhao, H., Liu, F., Xie, W., Zhou, T.C., OuYang, J., Jin, L., Li, H., Zhao, C.Y., Zhang, L., Wei, J., Zhang, Y.P., Li, C.P.: Ultrasensitive supersandwich-type electrochemical sensor for SARS-CoV-2 from the infected COVID-19 patients using a smartphone. Sens. Actuators, B Chem. **327**, 128899 (2021). https://doi.org/10.1016/j.snb.2020.128899
62. Ma, Y., Shen, X., Zeng, Q., Wang, H., Wang, L.: Talanta A multi-walled carbon nanotubes based molecularly imprinted polymers electrochemical sensor for the sensitive determination of HIV-p24 **164**, 121–127 (2017). https://doi.org/10.1016/j.talanta.2016.11.043.
63. Palomar, Q., Gondran, C., Marks, R., Cosnier, S., Holzinger, M.: Impedimetric quantification of anti-dengue antibodies using functional carbon nanotube deposits validated with blood plasma assays. Electrochim. Acta. **274**, 84–90 (2018). https://doi.org/10.1016/j.electacta.2018.04.099
64. Zhu, S., Song, Y., Zhao, X., Shao, J., Zhang, J., Yang, B.: The photoluminescence mechanism in carbon dots (graphene quantum dots, carbon nanodots, and polymer dots): Current state and future perspective **8**, 355–381 (2015). https://doi.org/10.1007/s12274-014-0644-3
65. Ahmed, S.R., Mogus, J., Chand, R., Nagy, E., Neethirajan, S.: Optoelectronic fowl adenovirus detection based on local electric field enhancement on graphene quantum dots and gold nanobundle hybrid. Biosens. Bioelectron. **103**, 45–53 (2018). https://doi.org/10.1016/j.bios.2017.12.028
66. Qaddare, S.H., Salimi, A.: crossmark. Biosens. Bioelectron. **89**, 773–780 (2017). https://doi.org/10.1016/j.bios.2016.10.033
67. Lee, J., Takemura, K., Park, E.Y.: Plasmonic/magnetic graphene-based magnetofluoro-immunosensing platform for virus detection. Sensors Actuators, B Chem. **276**, 254–261 (2018). https://doi.org/10.1016/j.snb.2018.08.124
68. Jin, S., Poudyal, S., Marinero, E.E., Kuhn, R.J., Stanciu, L.A.: Electrochimica Acta Impedimetric Dengue Biosensor based on Functionalized Graphene Oxide Wrapped Silica Particles. Electrochim. Acta. **194**, 422–430 (2016). https://doi.org/10.1016/j.electacta.2016.02.116
69. Chan, C., Shi, J., Fan, Y., Yang, M.: A microfluidic flow-through chip integrated with reduced graphene oxide transistor for influenza virus gene detection. Sens. Actuators, B Chem. **251**, 927–933 (2017). https://doi.org/10.1016/j.snb.2017.05.147
70. Negahdari, B., Darvishi, M., Saeedi, A.A.: Gold nanoparticles and hepatitis B virus. Artif. Cells, Nanomedicine Biotechnol. **47**, 469–474 (2019). https://doi.org/10.1080/21691401.2018.1546185
71. Nguyen, Q.H., Il Kim, M.: Since January 2020 Elsevier has created a COVID-19 resource centre with free information in English and Mandarin on the novel coronavirus COVID- 19. The COVID-19 resource centre is hosted on Elsevier Connect, the company's s public news and information (2020)
72. Teengam, P., Siangproh, W., Tuantranont, A., Vilaivan, T.: Multiplex Paper-Based Colorimetric DNA Sensor Using Pyrrolidinyl Peptide Nucleic Acid-Induced AgNPs Aggregation for Detecting MERS-CoV, MTB, and HPV Oligonucleotides, (2017). https://doi.org/10.1021/acs.analchem.7b00255
73. Akbari Hasanjani, H.R., Zarei, K.: DNA/Au-Pt bimetallic nanoparticles/graphene oxide-chitosan composites modified pencil graphite electrode used as an electrochemical biosensor for sub-picomolar detection of anti-HIV drug zidovudine. Microchem. J. **164**, 106005 (2021). https://doi.org/10.1016/j.microc.2021.106005.
74. Tian, J., Liang, Z., Hu, O., He, Q., Sun, D., Chen, Z.: An electrochemical dual-aptamer biosensor based on metal-organic frameworks MIL-53 decorated with Au@Pt nanoparticles and enzymes for detection of COVID-19 nucleocapsid protein. Electrochim. Acta. **387**, 138553 (2021). https://doi.org/10.1016/j.electacta.2021.138553
75. Draz, M.S., Vasan, A., Muthupandian, A., Kanakasabapathy, M.K., Thirumalaraju, P., Sreeram, A., Krishnakumar, S., Yogesh, V., Lin, W., Yu, X.G., Chung, R.T., Shafiee, H.: Virus detection using nanoparticles and deep neural network-enabled smartphone system. Sci. Adv. **6**, 1–13 (2020). https://doi.org/10.1126/sciadv.abd5354
76. Zou, R., Zhang, F., Chen, C., Cai, C.: Sensors and Actuators B : Chemical DNA-programming multicolor silver nanoclusters for sensitively simultaneous detection of two HIV DNAs. Sens. Actuators B. Chem. **296**, 126608 (2019). https://doi.org/10.1016/j.snb.2019.05.085

77. Kurdekar, A.D., Chunduri, L.A.A., Chelli, S.M., Haleyurgirisetty, M.K., Bulagonda, E.P., Zheng, J., Hewlett, I.K., Kamisetti, V.: RSC Advances Fluorescent Silver Nanoparticle Based Highly Sensitive Immunoassay for Early Detection of HIV, pp. 19863–19877 (2017). https://doi.org/10.1039/c6ra28737a
78. Xu, S., Ouyang, W., Xie, P., Lin, Y., Qiu, B., Lin, Z., Chen, G., Guo, L., Xu, S., Ouyang, W., Xie, P., Lin, Y., Qiu, B., Lin, Z., Chen, G.: Highly uniform gold nanobipyramids for ultrasensitive colorimetric detection of influenza virus Highly uniform gold nanobipyramids for ultrasensitive colorimetric detection of influenza virus (2016). https://doi.org/10.1021/acs.analchem.6b03711
79. Negahdari, B., Darvishi, M., Saeedi, A.A.: Gold nanoparticles and hepatitis B virus. Artif. Cells, Nanomedicine Biotechnol. **47**, 455–461 (2019). https://doi.org/10.1080/21691401.2018.1553786
80. Li, Z., Yi, Y., Luo, X., Xiong, N., Liu, Y., Li, S., Sun, R., Wang, Y., Hu, B., Chen, W., Zhang, Y., Wang, J., Huang, B., Lin, Y., Yang, J., Cai, W., Wang, X., Cheng, J., Chen, Z., Sun, K., Pan, W., Zhan, Z., Chen, L., Ye, F.: Development and clinical application of a rapid IgM—IgG combined antibody test for SARS—CoV—2 infection diagnosis, (n.d.). https://doi.org/10.1002/jmv.25727.
81. Zhang, Q., Li, L., Qiao, Z., Lei, C., Xie, Q., Yao, S., Li, Y., Ying, Y.: Electrochemical Conversion of Fe3O4 Magnetic Nanoparticles to Electroactive Prussian Blue Analogues for Novel Self-Sacrificial-Label Biosensing of Avian Influenza Virus H5N1 Electrochemical Conversion of Fe 3 O 4 Magnetic Nanoparticles to Electroactive P (2017). https://doi.org/10.1021/acs.analchem.7b02784.
82. Wang, Z., Zheng, Z., Hu, H., Zhou, Q., Liu, W., Li, X., Liu, Z., Wang, Y., Ma, Y.: A point-of-care selenium nanoparticle-based test for the combined detection of anti-SARS-CoV-2 IgM and IgG in human serum and blood. Lab Chip. **20**, 4255–4261 (2020). https://doi.org/10.1039/d0lc00828a
83. Liang, S., Qi, L., Zhang, R., Jin, M., Zhang, Z.: Sensors and Actuators B : Chemical Ratiometric fluorescence biosensor based on CdTe quantum and carbon dots for double strand DNA detection. Sensors Actuators B. Chem. **244**, 585–590 (2017). https://doi.org/10.1016/j.snb.2017.01.032
84. Deng, X., Wang, C., Gao, Y., Li, J., Wen, W., Zhang, X., Wang, S.: Biosensors and Bioelectronics Applying strand displacement amplification to quantum dots-based fluorescent lateral flow assay strips for HIV-DNA detection. Biosens. Bioelectron. **105**, 211–217 (2018). https://doi.org/10.1016/j.bios.2018.01.039
85. Ahmed, S.R., Kang, S.W., Oh, S., Lee, J., Neethirajan, S.: Chiral zirconium quantum dots: A new class of nanocrystals for optical detection of coronavirus. Heliyon. **4**, e00766 (2018). https://doi.org/10.1016/j.heliyon.2018.e00766

Carbon Quantum Dots

Thiago Leandro Oliveira and Annelise Kopp Alves

Abstract Carbon Quantum Dots (CQD's) are spherical particles formed by piled graphene fragments with particles sized less than 10 nm and functional groups on the surface. They can create high-performance nanocompounds that can be attained from many natural carbon sources. Due to this reason, many studies seek to produce CQD's via simple synthesis methods using low-cost precursors, such as residual organics like coffee powder. In this chapter, the synthesis methods of CQD's were summarized, and their luminescence mechanism was analyzed.

Keywords Carbon quantum dots · Photoluminescent properties · Nanoparticles · Natural carbon sources

Abbreviations

C60	Fullerene
CNT	Carbon Nanotube
rGO	Reduced Graphene Oxide
SWNT	Carbon Single-Walled Nanotube
CQD's	Carbon Quantum Dots
QD	Quantum Dots
NC	Nanocrystalline Semiconductor
UV	Ultraviolet
PL	Photoluminescence
RQ	Quantum Rate
QY	Quantum Yeld
CNP	Carbon Nanoparticle
PLQY	Photoluminescence Quantum Yield
EO	Electrochemical Oxidation

T. L. Oliveira (✉) · A. K. Alves
Universidade Federal do Rio Grande do Sul, Porto Alegre, Brazil

A. Kopp Alves (ed.), *Technological Applications of Nanomaterials*,
Engineering Materials, https://doi.org/10.1007/978-3-030-86901-4_4

1 Introduction

The last decade of the twentieth century was marked by the emergence of a new science area that disclosed a new perspective for materials engineering: nanotechnology. In brief, the basic principle of nanotechnology is based on the fact that the material properties depend not only on its composition and structure but also on its size and shape [1]. Then, to be considered nanotechnological, these new materials must have at least one of their dimensions, range 1–100 nm [2].

Amongst the nanomaterials, perhaps the most promising and known are carbon-based materials. Carbon atoms stand out for the exceptional ability to participate in consistent covalent bonds in different hybridization states (sp, sp^2, sp^3), thus forming a complex and varied range of materials [3]. Figure 1 represents some carbon-based nanomaterials elements.

In 2004, in an experiment involving the purification of carbon single-walled nanotube (SWNT) by electrophoresis, scientists reported the split of these nanotubes into two other material species. However, the first one was similar to the SWNT in smaller scales and, afterward, called short tubular carbon. The second species evidenced a nanoparticle format, having about 18 nm diameter and so, named Carbon Quantum Dots (CQD) [5].

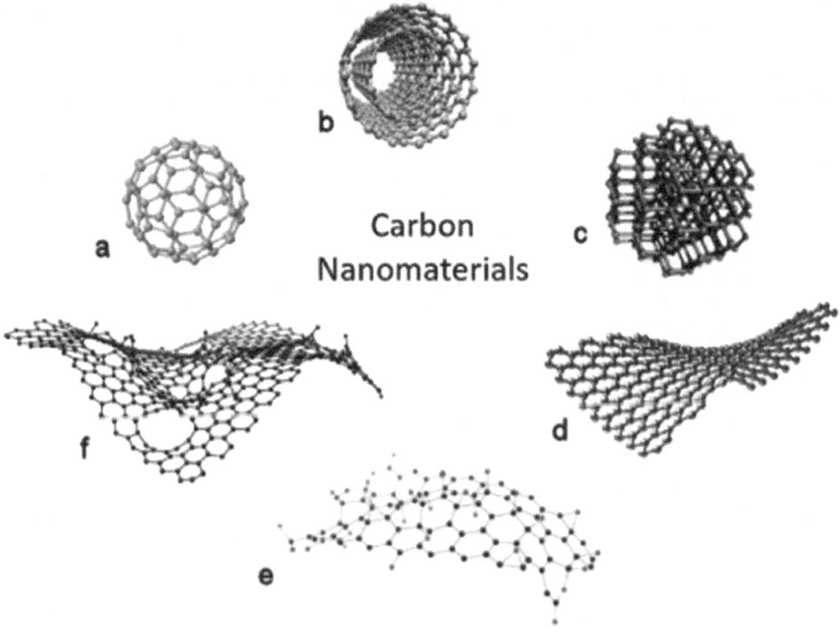

Fig. 1 Carbon nanomaterials: **a** Fullerene (C60); **b** Carbon Nanotube (CNT); **c** Nanodiamond; **d** Graphene; **e** Graphene Oxide; **f** reduced Graphene Oxide (rGO). Reprinted with permission from [4]

In general, CQD's are spherical particles formed by piled graphene fragments with particles sized less than 10 nm and functional groups on the surface. CQD's have eccentric properties. They are high-performance nanocompounds that can be attained from several natural carbon sources. Moreover, the high quantum yield and regulating emission wavelength, which are qualities found in traditional quantum dots (QD's), CQD's have many other excellent characteristics, including image constancy, low cytotoxicity, biocompatibility, easy surface modification, and high inert chemical [6].

There are several synthetic routes to achieve CQD's. They can be classified as physical methods, called top-down, and chemical methods, known as bottom-up. The chemical methods comprise electrochemical synthesis, combustion, acid oxidation, hydrothermal, thermal pyrolysis, destruction of fullerenes, carbon nanotubes, etc. The physical methods include arc flash discharge, laser ablation, and plasma treatment [7].

Most studies have searched for CQD's via simple synthesis methods using low-cost precursors, set the abundance of existing carbon sources such as coffee powder, watermelon rind, eggshell membrane, orange peel, ruminant manure, hair, etc. In this chapter, the synthetic methods of CQD's were summarized, and their luminescence mechanism was analyzed.

2 Carbon

Carbon is an essential element for all living species, and it is one of the versatile and most erratic elements already found. Since its discovery in the nineteenth century, studies regarding this element's characteristics have endlessly tried amplifying their knowledge about it [8].

Carbon has several allotropes and can connect with a choice of elements, making it an element found in over a million compounds [9].

The electron configuration of carbon confers a unique characteristic to its atoms of participating in covalent bonds and other carbon atoms in different hybridization states (sp, sp^2, and sp^3), that is, single, double and triple bonds. This property supports the immense importance of organic chemistry and biochemistry. Two centuries ago, it was first shown that carbon is present in organic molecules and biomolecules and natural carbon materials such as the various types of amorphous carbon, diamond, and graphite [3].

3 Carbon Nanoparticles

Carbon nanostructures can be divided into two groups that vary based on covalent bonds that connect their carbon atoms.

The first group involves graphene nanostructures, mainly composed of sp^2 carbon atoms, densely filled into a honeycomb hexagonal crystal structure. However, they may also contain some sp^3 carbon atoms at their imperfection or edges. This group consists of graphene, graphene nanosheets, CNTs, nanohorns, onion-type carbon nanospheres, and carbon quantum dots [3].

The second group of carbon nanostructures holds sp^3 and sp^2 bonds in various proportions and has mixtures of amorphous and graphitic regions or consists mainly of sp^3 carbon atoms. Nonetheless, the nanodiamond is the only known member of this group, and some types of CQD's, with non-graphitic structures, may be also considered members [3].

Carbon structures can be classified according to their morphological characteristics. The first category in such a classification scheme would enclose nanostructures with internal empty spaces such as fullerene, carbon nanotubes, and nanohorns. Nanostructures without internal spaces equally nanodiamonds, CQD's, and graphene would belong to the second category in this design.

Finally, it is possible to classify these nanostructures according to the dimensionality of the structures. This design considers 0D carbon structures such as fullerenes, CQD's, and nanodiamonds; 1D nanoallotropes like CNT's and carbon nanofibers; 2D nanoallotropes such as graphene and 3D nanoribbons and nanoallotropes such as SWCNT and CWCNT. In Fig. 2, we have the representation of carbon allotropes and their properties of hybridization, crystal system, dimension, specific surface area, and density.

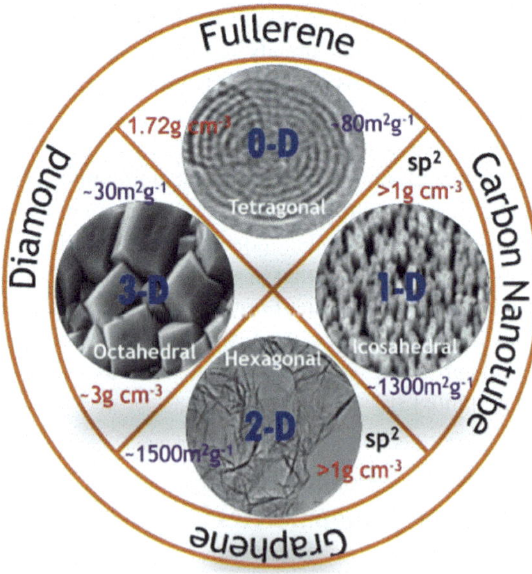

Fig. 2 Representation of carbon allotropes and their properties of hybridization (black), crystal system (white), dimension (blue), specific surface area (lavender), and density (red). Reprinted with permission from [10]

4 Quantum Dots and Carbon Quantum Dots

4.1 Quantum Dots

Quantum Dots (QD's), also known as nanocrystalline semiconductors, after excitation by electromagnetic radiation, produce an electron–hole pair identified as an exciton. This exciton undergoes strong quantum confinement in the three space dimensions, which gives the QD's some interesting optical properties, such as the strong dependence of emission due to the particle size. One of the most important properties of QD's is photoluminescence. Below, Fig. 3 will represent the emission dependence with the nanoparticle size.

Currently, QD's are used in optoelectronic devices, photovoltaic and biomedical applications. However, most synthetic routes for QD's use heavy metals such as cadmium. Consequently, it can represent severe limitations for its applications due to the known toxicity and potential environmental risk associated with this metal. Therefore, the search for replacers to QD's containing these metals but with similar properties is essential to match the applications of these materials.

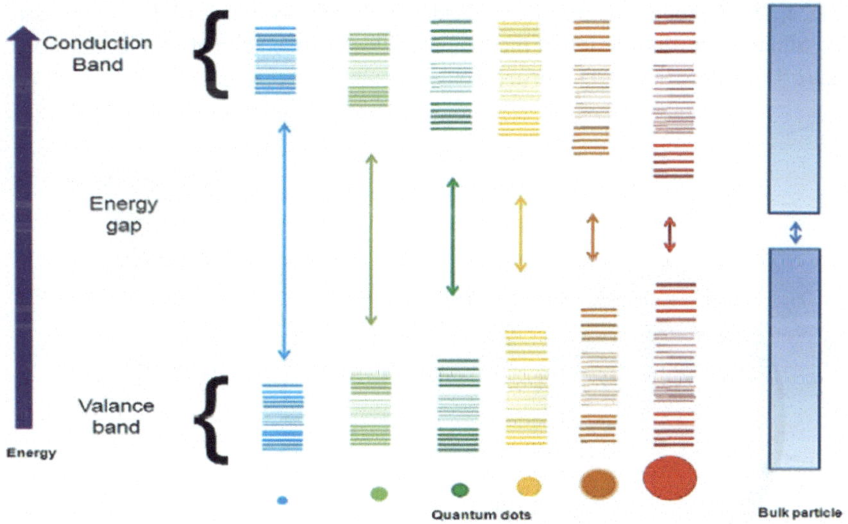

Fig. 3 Illustration of the effect of forbidden band energy variation as a result of the size of the nanoparticles. Adapted with permission from [11]

4.2 Carbon Quantum Dots

The CQD's, Carbon Quantum Dots, C-dots, Carbon Dots, or Carbon Nanodots consist of a zero-dimensional (0D) nanoparticles class sized lower than 10 nm, which bandgap electronic structures are highly influenced by the quantum imprisonment effect (Fig. 3) [12].

CQD's have been compared to QD's (Quantum Dots) as they have similar optical properties such as photoluminescence and intense dependence on emission with their particle size. Nevertheless, CQD's are a viable alternative to traditional QD's for using low-cost and less widespread synthetic courses, long-term colloidal stability, essential loads, and low environmental and biological toxicity [13].

They are commonly used for fluorescent carbogenic materials with an outer layer composed of a carboxyl or other chemical functional group and a graphite center, eventually containing covalent connections between oxygen and nitrogen atoms.

CQD's features has been extensively examined by Fourier transform infrared spectroscopy (FTIR), nuclear magnetic resonance (NMR), and X-ray photoelectron spectroscopy (XPS) [14]. The set of available results reveals that the CQD's are characteristically quasi-spherical nanoparticles and mainly consist of sp^2 carbon with graphite-compatible interline spacing, which may contain polar clusters on the surface. These groups are introduced during or after the synthesis of these nanoparticles, allowing the dissolution of the nanoparticles in polar solvents, including water. Figure 4 illustrate the CQD's structure.

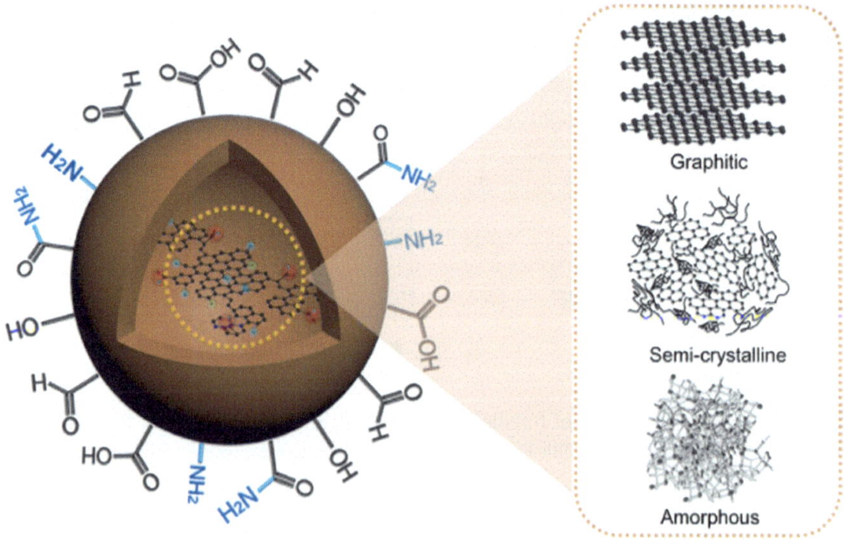

Fig. 4 The chemical structure of CDs includes a core structure of graphitic (i.e., graphene quantum dots), semi-crystalline (i.e., carbon quantum dots or carbon nanodots), or amorphous features (i.e., polymer dots), and a surface decorated with various functional groups. Adapted from [15]

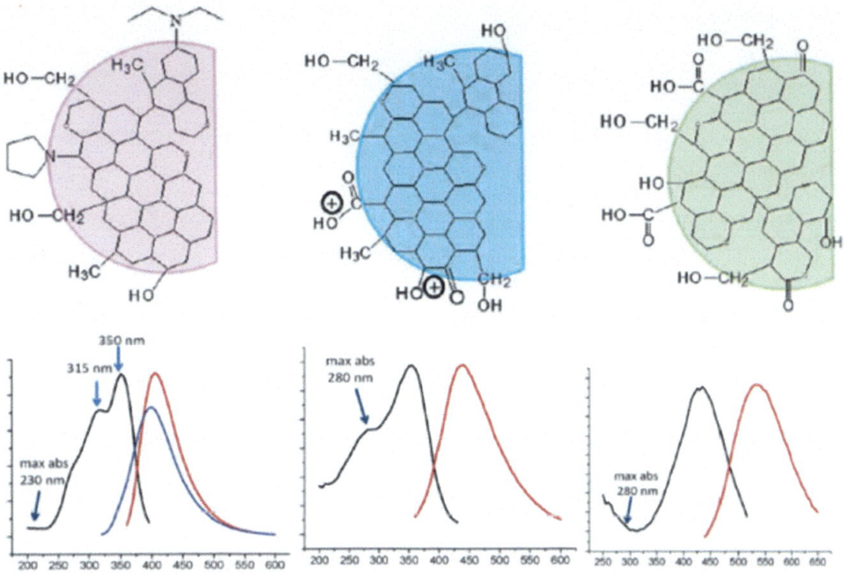

Fig. 5 Schematic representation of three types of C-dots with different surface groups and their typical fluorescence excitation and emission spectra. The "violet" C-dots **a** is with nitrogen- and oxygen-containing groups, "blue" **b** with oxygen-containing groups and cations, and "green" **c** C-dots contain oxygen groups only. Reprinted from [16], open access with MDPI permission

Regarding the photophysical properties of CQD's, these materials have an intense absorption band in the ultraviolet (UV), and the photoluminescence (PL) spectrum depends on the regulating emission wavelength. The origin of the CQD's PL is still not fully understood. Still, it is probably related to the quantum internment effect due to the nanometer size of the particles or different emissive traps that adhere to the CQD's surface. Figure 5 compares some CQD's structures with other functional groups on the surface and their luminescent coloration [14].

C, H, and O are the elementary creators of CQD's in varying concentrations. The heteroatom's introduction has been studied to regulate the luminescent emission. For instance, the CQD's quantum rate (QR) may be significantly enhanced through the incorporation of N; meanwhile, the incorporation of S may increase the emission bandwidth [14].

5 CQD's Synthesis Methods

Last decade, a variety of techniques was proposed for CQD's production. These efforts not only synthesized the CQD's but also improved their properties and simplified the production process. Commonly, CQD's preparation can be done within two main courses: the top-down and bottom-up methods [17].

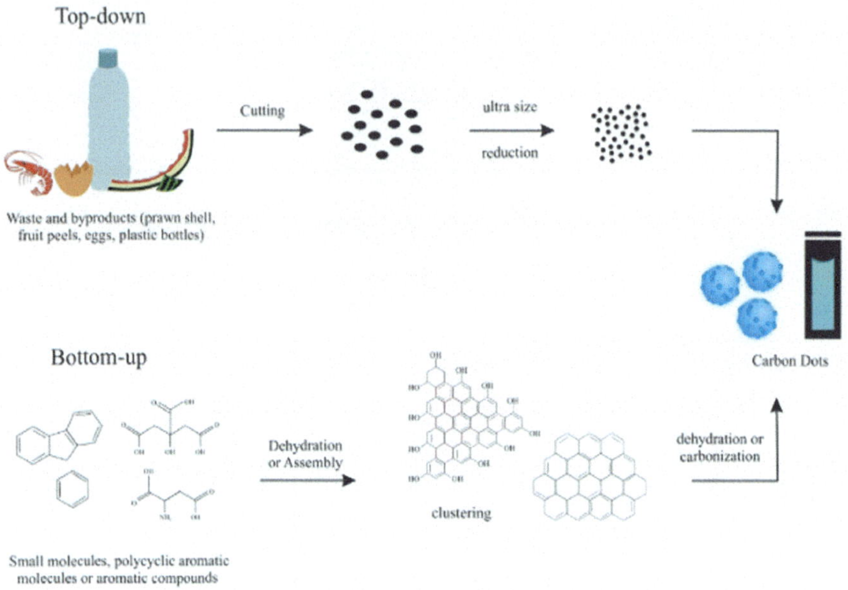

Fig. 6 Schematic representation of top-down and bottom-up synthesis approaches from waste and by-products. Reprinted with permission from [19]

Top-down method approaches consist of electric arc discharges, laser ablation, and electrochemical oxidation, where CQD's are formed or "broken" from a larger carbon structure such as nanotubes, graphene, graphite, and activated carbon. Unfortunately, through the CQD's top-down synthesis, harsh experimental conditions such as high voltage discharge, expensive operating steps and equipment are commonly used, limiting their practical application.

The bottom-up method consists, for instance, of ignition/thermal, hydrothermal, or microwave processes in which CQD's are formed from molecular precursors [18]. They have the advantages of price-performance, easy operation, and simple equipment requirements, and therefore have been widely used in the synthesis of QCD's. Figure 6 presents the schematic representation of top-down and bottom-up synthesis.

5.1 *CQD's Synthesis from Biomass Residues*

Biomass consists of all organic material originated from plants or animals used to produce energy. It is a plentiful, complex, and biodegradable carbon source, oxygen, and nitrogen (sulfur is found in few cases) mainly composed of cellulose, hemicellulose, lignin, ash and proteins. Hence, considering the biomass of plant origin, its cellulose composition varies from 30 to 60%, hemicellulose is between 20 and 40%, and lignin is from 15 to 25% [20].

As an organic, ecologically sustainable, abundant waste and a natural source of carbon, biomass is an excellent raw material for the production of CQD's. However, the most significant part of organic materials used to compose the biomass is treated as waste and mistakenly rejected. Such a way has interested in using biomass as a raw material to produce CQD's grown exponentially. The main methods for CQD's production from biomass residues will be treated along following topics.

5.2 *Pyrolysis*

The pyrolysis synthesis or carbonization is a straightforward method of thermal association, which takes place in an inert or oxidizing atmosphere, and a further purification treatment is wanted. A carbon source's organic material is gradually converted into CQD's through heating, dehydration, degradation, and carbonization at high temperatures [6].

This method usually requires the intense use of alkali or acid to break down the carbon source precursor molecules into nanoscale particles. Many biomass types may be used for CQD's production via pyrolysis, including watermelon rind [21], peanut shell [22], lychee seeds [23], etc. The properties of CQD's reaches can be synchronized by changing the pyrolysis conditions such as temperature, time, and pH value of the reaction systems [24].

Zhou et al. [21] performed the pyrolysis of watermelon rind at 220 °C for 2 h in an oxidizing atmosphere; the process was carried out using a muffle furnace to achieve CQD's. After the carbonization process, the sample was separate in water, sonicated, filtered, and centrifuged, producing a suspension of CQD's with a bright blue light. These CQD's were used in bio-imaging (Fig. 7a).

Xue et al. [23] carbonized lychee seeds in a ceramic crucible at 300 °C for 2 h. Then, the sample was separated in water, sonicated, and filtered, forming a suspension, also blue (Fig. 7c).

Xue et al. [22] used peanut shells to obtain N-doped CQD's at a 9,91% quantum yield, which have well steadiness, noteworthy photobleaching resistance, high tolerance to large pH fluctuations and ionic strength. At first, the peanut shells were placed in a ceramic crucible and carbonized at a 250 °C temperature for 2 h. After cooling to room temperature, the dark black products were mechanically ground into fine powders. Then, 0.1 g of each achieved sample was dispersed by ultrasound in 10 mL of water to get to a homogeneous solution. Finally, the CQD's were collected by removing larger particles through filtration (Fig. 7b).

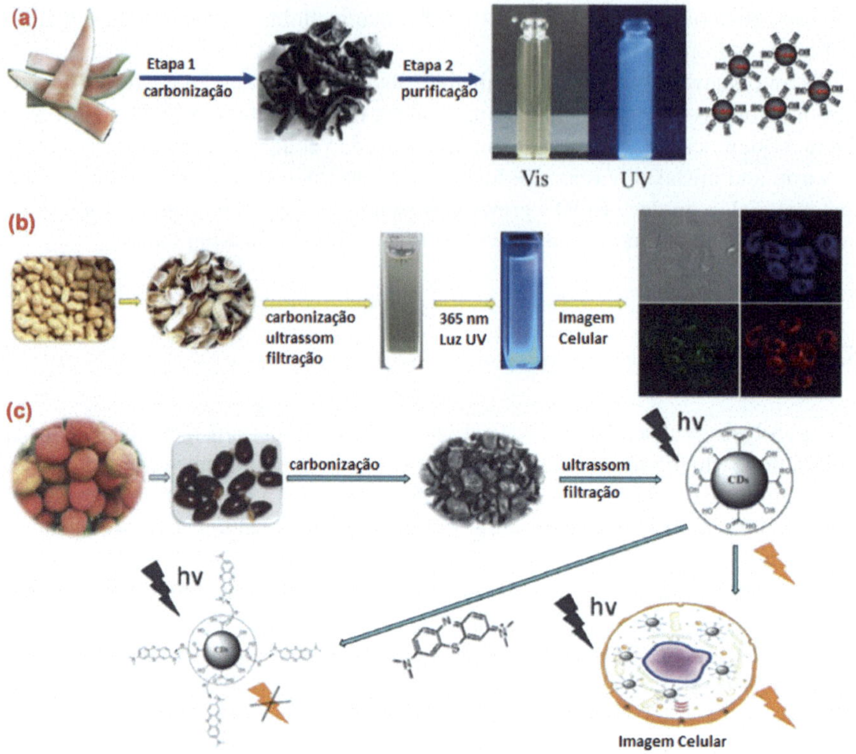

Fig. 7 CQD's synthesis from biomass via pyrolysis **a** CQD's synthesis dissolved in water from watermelon rind [21]. **b** CQD's synthesis from lychee seeds [23]. **c** CQD's synthesis and applications from peanut shells [22]. Reprinted with MDPI permission from [6]

5.3 Solvothermal Method

One of the most reported synthesis methods in the literature is the solvothermal treatment called hydrothermal treatment. This technique is widely used because it is considered a cheap, non-toxic, and simple technique. The reaction takes place inside a sealed autoclave reactor, in which there is a solution containing the solvent, usually water, and the organic precursor. At some stage in solvothermal treatment, dehydration, polymerization, and nucleation reactions occur, as observed by Yang et al. [25]. Through this approach, it is worth mentioning that it is easier to control the size of the synthesized nanoparticles, and consequently, the fluorescence of the CQD's, observing better colloidal stability of the material in water as well. In this type of process, it is possible to have CQD's with different characteristics, from the temperature used and the pressure in the container.

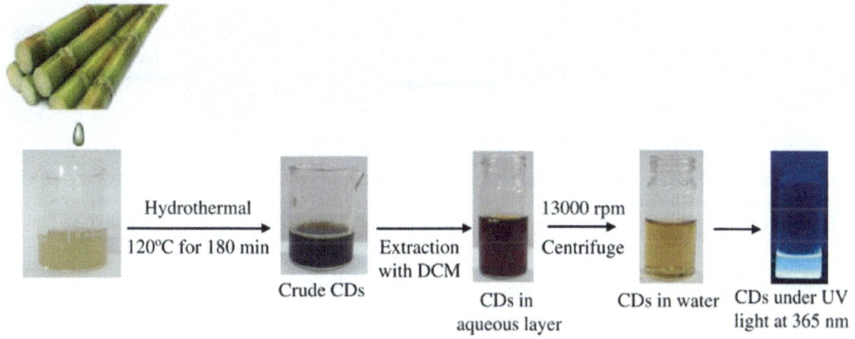

Fig. 8 Schematic representation for the formation of carbon dots through the hydrothermal treatment of *Saccharum officinarum* juice. Reprinted with permission from [27]

Sahu et al. have prepared highly photoluminescent CQD's with a quantum yield (QY) of 26% using orange juice solvothermal treatment followed by centrifugation. These CQD's sized 1.5–4.5 nm were applied in bio-imaging due to their high photostability and low toxicity [26].

Sugarcane juice (*Saccharumofficinarum*) was also used for the hydrothermal synthesis of CQD's. Metha et al. mixed the sugarcane juice with ethanol and placed it in a sealed container autoclave at 120°C heated for 180 min. When removed from the autoclave, the solution displayed a dark brown color being undergone into a purification and withdrawal process which, when excited, showed a blue luminescence [27] (Fig. 8).

5.4 *Microwaves*

The microwave-assisted method is a widely used process to carbonize organic substances into CQD's under microwave radiation directly. Due to its efficiency, simplicity in device and operation, the microwave-assisted method is an inexpensive approach with a competitive advantage for producing large quantities of fluorescent CQD's.

In comparison to other methods, it is more convenient and faster to heat the carbon precursors this way, which simplifies the synthesis process so that the CQD's are easily attained in a few minutes with better results [6].

For example, a microwave-assisted method for the synthesis of carbon nanoparticles was achieved starting from microwave heating of 1-butyl-3-methylimidazoliumtetrafluo-roborate ([Bmim] BF4), described by Xiao [28]. This simple process for large-scale preparation of CQD's dissolved the precursor in water under stirring to form a clear solution. Then, the solution was heated in a microwave oven (700 W) followed by centrifugation to remove suspended solid impurities, resulting in a yellow–brown solution of CQD's (Fig. 9). The mean particle size of

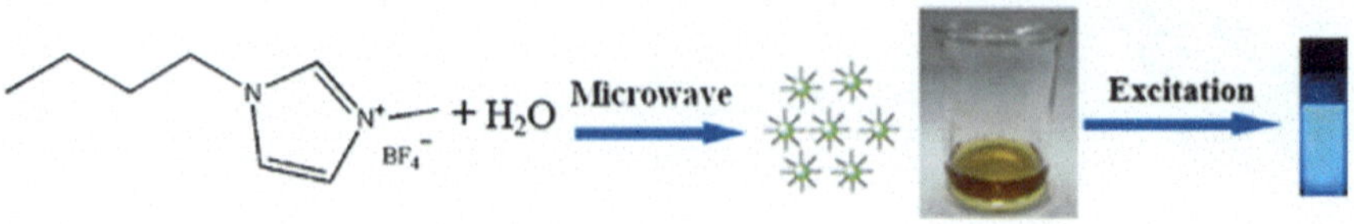

Fig. 9 A schematic illustration of the preparation procedure of CQD's by microwave pyrolysis. Reprinted with permission from [28]

the material was 4.25 ± 1.85 nm, having a uniform spherical shape. Thus, these nanoparticles can be used in biomedical and environmental analyzes in place of toxic metal-based semiconductor QDs. Furthermore, because of its low cytotoxicity, bright fluorescence, and excellent biocompatibility, this CQD's can be used to detect residual quercetin levels.

Other approach to synthesized CQD's using microwave-assisted synthesis from folic acid molecules as carbon and nitrogen sources was explored by Guan et al. [29]. In summary, a mixture of 15 mg of folic acid dissolved in 3 mL of diethylene glycol was initially prepared and after placed in a 750 W domestic microwave oven for 40 s. A reddish-brown suspension was obtained and submitted to pure water dialysis for 3 days. After post-treatment, luminescent carbon nitride nanoparticles with an average size of around 4.51 nm were synthesized. With less than 360 nm movement, the quantum yield of carbon nitride nanoparticles was estimated to be 18.9%. Interestingly, the emission peak position held almost invariant (at 460 nm) when illuminated at different wavelengths (320–420 nm).

5.5 *Ultrasonic*

The synthesis of CQD's via ultrasonic stands out concerning the other methods covered by its low cost and easy operation. CQD's can be obtained through the ultrasonic treatment of solvents mixture and carbon sources. The properties of CQD's can be regulated simply by adapting experimental conditions, such as ultrasonic power, reaction time, the proportion of solvents, and carbon sources. Park [30] proposed a simple method based on ultrasonic treatment for the large-scale synthesis of water-soluble CQD's from a carbon source derived from food residues and achieved 120 g of CQD's from 100 kg of food waste mixtures. The achieved CQD's displayed high solubility in water, high photostability, superior photoluminescence, and low cytotoxicity for in vitro bio-imaging. Furthermore, the sub-products produced in the synthesis of CQD's from the source derived from food residues can lead to seed germination and plant growth.

5.6 *Other Methods from Top-Down to CQD's Synthesis*

5.6.1 Electric Arc

It is, perhaps, the first used method to reach CQD's. As stated before, Xu et al. managed to isolate, by electric arc discharge, a fluorescent carbon-based nanomaterial [5].

This strategy can synthesize CQD's from carbon nanotubes or graphene sheets as raw material. First, CQD's are synthesized after the discharge of an electric arc in a controlled atmosphere, and then, the sediments are oxidized, usually with HNO_3, to induce carboxyl groups functionalization. The resulting material is extracted by an alkaline solution (NaOH), with defined pH, and a stable suspension of dark color is attained. Next, gel electrophoresis process is applied to purify the extracted material. From electrophoresis, it is possible to separate the CQD's by the sizes of their energy bands, which present different colors when illuminated by a specific wavelength [31].

5.6.2 Laser Ablation

Laser ablation is a CQD's production method that uses a laser beam with a specific wavelength to remove from a target surface made of carbon (C-target) the molecules necessary to form CQD's.

Sun et al. [32] reported the use of laser ablation method with a Q-switched Nd: YAG laser (1064 nm, 10 Hz) to prepare the CQD's using argon as the carrier gas in water vapor. However, the fluorescent carbon particles that resulted even after oxidation by nitric acid did not exhibit any detectable photoluminescent properties. Considerable surface passivation in acid oxidized the CQD's and a Photoluminescence Quantum Yield (PLQY) of up to 10% was achieved. This suggests that CQD's produced using this method requires several steps and complex conditions to display competent photoluminescence [17].

5.6.3 Electrochemical Oxidation

The Electrochemical oxidation (EO) is a CQD's synthesis method that uses an anode and a cathode (one of them being made of carbon) in an oxidative environment so that the product of this oxidation–reduction is CQD'S.

Ming et al. [33] applied a continuous voltage between two graphite electrodes immersed in deionized water. When using the potential difference to the electrodes, one of them becomes the cathode (positive pole) and the other the anode (negative pole), thus causing reduction and oxidation reactions. In such a way, when the graphite electrode starts to undergo oxidation, the chemical bonds between the carbon atoms are broken, and its structure is "exfoliated", disintegrating gradually leading to the formation of quantum dots of carbon in water.

Lu et al. [34] used high purity graphite rods as anode and platinum wire as cathode, in an aqueous solution. The application of static potentials initiated the oxidation of carbonic materials. The oxidation process was carried out, and as a result, CQD's sized 6-8 nm with QY of 2.8–5.2% were synthesized.

6 Main Characteristics Affecting the CQD's Properties

Photoluminescence (PL), and its applications, is the essential characteristic of CQD's. So far, CQD's of several sizes and PL wavelengths emission ranging from ultraviolet (UV, 200–380 nm) to infrared (NIR, 750–2000 nm) have been prepared.

Nevertheless, the mechanism responsible for CQD's photoluminescence phenomenon hasn't been adequately understood yet. Graphite nanostructures such as short carbon nanotubes and CQDs synthesized from graphite are fluorescent. However, graphitic carbon is an electrical conductor and does not have a valence band like semiconductors do. Therefore, the luminescence of graphitic nanostructures cannot be explained by these theories. However, some ideas have been exposed in an attempt to justify such a phenomenon. The conclusion is that there are three main factors which led to the CQD's photoluminescence properties: the effect of quantum confinement, surface "defects," and surface passivation.

As CQD's are prepared using several elements through multiple approaches containing different components and complex structures, it is tough to compare the results available in the literature to formulate a unified theory. Since changing synthesis conditions is the main resource to regulate the properties of CQD's, the impact of these conditions must be carefully analyzed.

6.1 Quantum Confinement Effects

Quantum Confinement Effects occur when the almost continuous electronic level, close to the Fermi level, transfers electrons to detached energy levels when the particle size decreases to the nanoscale.

Therefore, nanomaterials, especially those having a particle size below 10 nm, show different optical properties than their block counterparts. Li et al. prepared a series of CQDs in an electrochemical method followed by chromatography column division. They found an emission dependent on the size of the nanoparticle, such as ultraviolet light at 1.2 nm, visible light at 1.5–3.0 nm, and infrared light close to 3.8 nm. As shown in Fig. 10, it was claimed that the gap decreased while particle size increased [35].

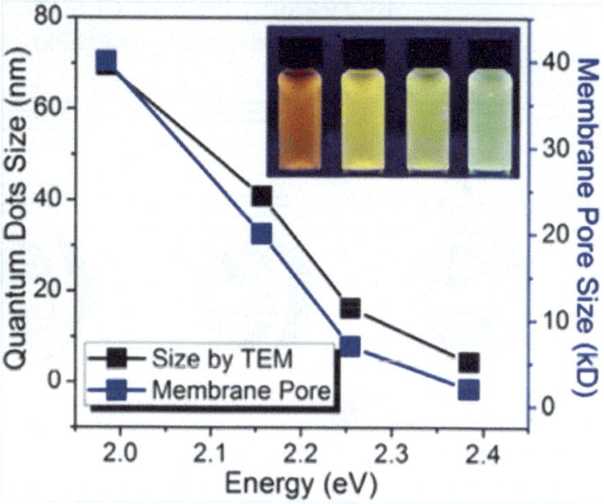

Fig. 10 Relationship between the optical bandgap and CQD size (from TEM) or membrane pore sizes used in the ultrafiltration. Adapted with permission from [36]

6.2 *Surface Defects*

Another fluorescence mechanism in CQD's occurs by their surface defect states. This type of defect refers to a boundary region or a spherical shell distinct from the carbon core region. The spheroidal region contains several chemical groups derived from sp^2 and sp^3 hybrid carbons, other surface functional groups, and pendant bonds. Due to the diversity and complexity of surface defects, the originating fluorescence is characterized by pleochroism and excitation-dependent luminescence. Surface defects are mainly generated by surface oxidation and can perform as a capture center for excitons, resulting in fluorescence.

The surface defect fluorescence is caused by the relaxation of radiation from the excited state to the ground state. The sp^3 and sp^2 hybrid carbon on the surface of CQD's and other weak defects can lead to multicolor emissions from their local electronic states. When the light of a specific wavelength illuminates the CQD's, the photons that energy satisfies the optical gap will transition and accumulate in adjacent surface defect traps later. Also, it will return to the ground state to emit visible light at different wavelengths. The greater the degree of CQD's surface oxidation, the more surface defects, and emission sites, resulting in the redshift of the wavelength (Fig. 11).

The surface state is not composed of an isolated chemical group but rather a center of fluorescence formed by the synergy of the carbon nucleus with coupled chemical groups. Among them, some groups are fluorescence activation state or absorption state under a certain condition. Functional groups have different energy levels and can produce several emission traps. The energy levels of functional groups may be allied to their ability to supply electrons. The stronger the ability of functional groups to provide electrons, the more energy they generate. The emission wavelength can be adapted by changing the chemical groups on CQD's surface.

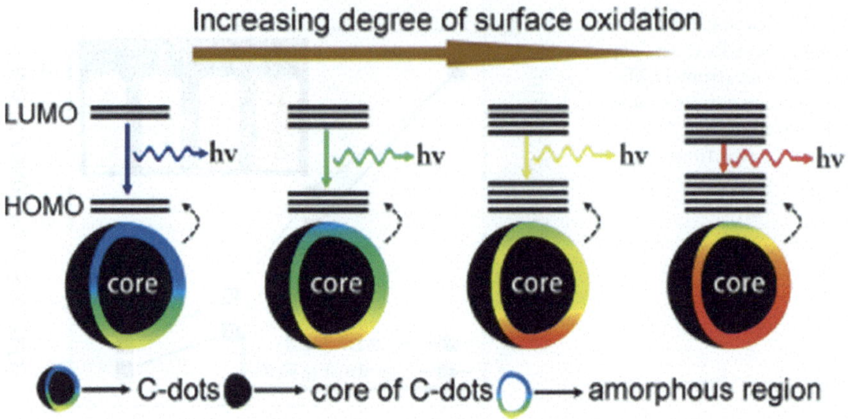

Fig. 11 Correlation between the different degrees of surface oxidation and the CQD's bandgap size. Reprinted with permission from [37]

6.3 Surface Passivation

The presence of functional groups on the surface seems to be a necessary condition for the fluorescence of CQD's since fluorescence is strongly affected by the pH of the medium. For instance, fluorescence intensity is conditioned by the ionization of functional groups on the surface of nanoparticles resulting from extreme pH.

Functional groups on the surface of CQD's can significantly affect the fluorescence properties of CQD's. Therefore, surface passivation or functionality plays an essential role in regulating the luminous properties of CQD's. Surface passivation treatment of CQD's can increase photoluminescence intensity, change the emission wavelength, narrow the fluorescence peak width, and improve water needlessly.

In addition, some other CQD's properties, such as selective assembling, can also be patterned to some extent by surface passivation. For instance, improved cell selectivity for staining nucleoli was achieved by simple amine passivation of CQD's synthesized from cow manure.

References

1. Zarbin, A.J.G., Oliveira, M.M.: Nanoestruturas de carbono (nanotubos, grafeno): Quo Vadis? Quim. Nova. **36**, 1533–1539 (2013). https://doi.org/10.1590/S0100-40422013001000009
2. Carneiro, S.V.: Pontos quânticos de carbono e nanopartículas de prata: potencial aplicação para sensoriamento de pesticidas. Universidade Federal Do Ceará Centro (2019). https://doi.org/10.1017/CBO9781107415324.004
3. Georgakilas, V., Perman, J.A., Tucek, J., Zboril, R.: Broad family of carbon nanoallotropes: classification, chemistry, and applications of fullerenes, carbon dots, nanotubes, graphene, nanodiamonds, and combined superstructures. Chem. Rev. **115**, 4744–4822 (2015). https://doi.org/10.1021/cr500304f

4. Giraud, L., Tourrette, A., Flahaut, E.: Carbon nanomaterials-based polymer-matrix nanocomposites for antimicrobial applications: A review. Carbon N. Y. **182**, 463–483 (2021). https://doi.org/10.1016/j.carbon.2021.06.002
5. Xu, X., Ray, R., Gu, Y., Ploehn, H.J., Gearheart, L., Raker, K., Scrivens, W.A.: Electrophoretic analysis and purification of fluorescent single-walled carbon nanotube fragments. J. Am. Chem. Soc. **126**, 12736–12737 (2004). https://doi.org/10.1021/ja040082h
6. Kang, C., Huang, Y., Yang, H., Yan, X.F., Chen, Z.P.: A review of carbon dots produced from biomass wastes. Nanomaterials. **10**, 1–24 (2020). https://doi.org/10.3390/nano10112316
7. Machado, C.E., Vieira, K.O., Ferrari, J.L., Schiavon, M.A.: Carbon quantum dots: chemical synthesis, properties and applications. Rev. Virtual Química. **7**, 1306–1346 (2015). https://doi.org/10.5935/1984-6835.20150073
8. Tian, P., Tang, L., Teng, K.S., Lau, S.P.: Graphene quantum dots from chemistry to applications. Mater. Today Chem. **10**, 221–258 (2018). https://doi.org/10.1016/j.mtchem.2018.09.007
9. Haley, M.M.: Carbon allotropes: On the road to carbyne. Nat. Chem. **2**, 912–913 (2010). https://doi.org/10.1038/nchem.884
10. Maduraiveeran, G., Jin, W.: Materials Science & Engineering B Carbon nanomaterials : Synthesis, properties and applications in electrochemical sensors and energy conversion systems. Mater. Sci. Eng. B. **272**, 115341 (2021). https://doi.org/10.1016/j.mseb.2021.115341
11. Sharma, S., Dutta, V., Singh, P., Raizada, P., Rahmani-Sani, A., Hosseini-Bandegharaei, A., Thakur, V.K.: Carbon quantum dot supported semiconductor photocatalysts for efficient degradation of organic pollutants in water: A review. J. Clean. Prod. **228**, 755–769 (2019). https://doi.org/10.1016/j.jclepro.2019.04.292
12. Yuan, T., Meng, T., He, P., Shi, Y., Li, Y., Li, X., Fan, L., Yang, S.: Carbon quantum dots: an emerging material for optoelectronic applications. J. Mater. Chem. C. **7**, 6820–6835 (2019). https://doi.org/10.1039/C9TC01730E
13. Dong, Y., Wang, R., Li, H., Shao, J., Chi, Y., Lin, X., Chen, G.: Polyamine-functionalized carbon quantum dots for chemical sensing. Carbon N. Y. **50**, 2810–2815 (2012). https://doi.org/10.1016/j.carbon.2012.02.046
14. Brito, J.B.: Pontos Quânticos em Matriz de Sílica Produzidos em Alta Pressão, Universidade Federal do Rio Grande do Sul (2017). https://www.lume.ufrgs.br/handle/10183/169589
15. Luo, H., Guo, Q., Szilágyi, P.Á., Jorge, A.B., Titirici, M.M.: Carbon dots in solar-to-hydrogen conversion. Trends Chem. **2**, 623–637 (2020). https://doi.org/10.1016/j.trechm.2020.04.007
16. Demchenko, A.: Excitons in carbonic nanostructures, C. J. Carbon Res. **5**, 71 (2019). https://doi.org/10.3390/c5040071
17. Xu, Q., Kuang, T., Liu, Y., Cai, L., Peng, X., SreenivasanSreeprasad, T., Zhao, P., Yu, Z., Li, N.: Heteroatom-doped carbon dots: synthesis, characterization, properties, photoluminescence mechanism and biological applications. J. Mater. Chem. B. **4**, 7204–7219 (2016). https://doi.org/10.1039/C6TB02131J
18. Baker, S.N., Baker, G.A.: Luminescent carbon nanodots: Emergent nanolights. Angew. Chemie Int. Ed. **49**, 6726–6744 (2010). https://doi.org/10.1002/anie.200906623
19. Wang, Y., Hu, A.: Carbon quantum dots: Synthesis, properties and applications. J. Mater. Chem. C. **2**, 6921–6939 (2014). https://doi.org/10.1039/c4tc00988f
20. Hosseini Koupaie, E., Dahadha, S., Bazyar Lakeh, A.A., Azizi, A., Elbeshbishy, E.: Enzymatic pretreatment of lignocellulosic biomass for enhanced biomethane production-A review. J. Environ. Manage. **233**, 774–784 (2019). https://doi.org/10.1016/j.jenvman.2018.09.106
21. Zhou, J., Sheng, Z., Han, H., Zou, M., Li, C.: Facile synthesis of fluorescent carbon dots using watermelon peel as a carbon source. Mater. Lett. **66**, 222–224 (2012). https://doi.org/10.1016/j.matlet.2011.08.081
22. Xue, M., Zhan, Z., Zou, M., Zhang, L., Zhao, S.: Green synthesis of stable and biocompatible fluorescent carbon dots from peanut shells for multicolor living cell imaging. New J. Chem. **40**, 1698–1703 (2016). https://doi.org/10.1039/c5nj02181b
23. Xue, M., Zou, M., Zhao, J., Zhan, Z., Zhao, S.: Green preparation of fluorescent carbon dots from lychee seeds and their application for the selective detection of methylene blue and imaging in living cells. J. Mater. Chem. B. **3**, 6783–6789 (2015). https://doi.org/10.1039/C5TB01073J

24. Zuo, P., Lu, X., Sun, Z., Guo, Y., He, H.: A review on syntheses, properties, characterization and bioanalytical applications of fluorescent carbon dots. Microchim. Acta. **183**, 519–542 (2016). https://doi.org/10.1007/s00604-015-1705-3
25. Yang, Z.C., Li, X., Wang, J.: Intrinsically fluorescent nitrogen-containing carbon nanoparticles synthesized by a hydrothermal process. Carbon N. Y. **49**, 5207–5212 (2011). https://doi.org/10.1016/j.carbon.2011.07.038
26. Sahu, S., Behera, B., Maiti, T.K., Mohapatra, S.: Simple one-step synthesis of highly luminescent carbon dots from orange juice: Application as excellent bio-imaging agents. Chem. Commun. **48**, 8835–8837 (2012). https://doi.org/10.1039/c2cc33796g
27. Mehta, V.N., Jha, S., Kailasa, S.K.: One-pot green synthesis of carbon dots by using Saccharum officinarum juice for fluorescent imaging of bacteria (Escherichia coli) and yeast (Saccharomyces cerevisiae) cells. Mater. Sci. Eng. C. **38**, 20–27 (2014). https://doi.org/10.1016/j.msec.2014.01.038
28. Xiao, D., Yuan, D., He, H., Gao, M.: Microwave assisted one-step green synthesis of fluorescent carbon nanoparticles from ionic liquids and their application as novel fluorescence probe for quercetin determination. J. Lumin. **140**, 120–125 (2013). https://doi.org/10.1016/j.jlumin.2013.02.032
29. Guan, W., Gu, W., Ye, L., Guo, C., Su, S., Xu, P., Xue, M.: Microwave-assisted polyol synthesis of carbon nitride dots from folic acid for cell imaging. Int. J. Nanomedicine. **9**, 5071–5078 (2014). https://doi.org/10.2147/IJN.S68575
30. Park, S.Y., Lee, H.U., Park, E.S., Lee, S.C., Lee, J.W., Jeong, S.W., Kim, C.H., Lee, Y.C., Huh, Y.S., Lee, J.: Photoluminescent green carbon nanodots from food-waste-derived sources: Large-scale synthesis, properties, and biomedical applications. ACS Appl. Mater. Interfaces. **6**, 3365–3370 (2014). https://doi.org/10.1021/am500159p
31. Singh, I., Arora, R., Dhiman, H., Pahwa, R.: Carbon quantum dots: Synthesis, characterization and biomedical applications. Turkish J. Pharm. Sci. **15**, 219–230 (2018). https://doi.org/10.4274/tjps.63497
32. Sun, X., Liu, Y., Niu, N., Chen, L.: Synthesis of molecularly imprinted fluorescent probe based on biomass-derived carbon quantum dots for detection of mesotrione. Anal. Bioanal. Chem. **411**, 5519–5530 (2019). https://doi.org/10.1007/s00216-019-01930-y
33. Ming, H., Ma, Z., Liu, Y., Pan, K., Yu, H., Wang, F., Kang, Z.: Large scale electrochemical synthesis of high quality carbon nanodots and their photocatalytic property. Dalt. Trans. 0–5 (2012). https://doi.org/10.1039/C2DT30985H
34. Lu, J., Yang, J., Wang, J., Lim, A., Wang, S., Loh, K.P.: One-Pot Synthesis of Fluorescent Carbon Graphene by the Exfoliation of Graphite in Ionic Liquids1. Lu, J., Yang, J., Wang, J., Lim, A., Wang, S., Loh, K.P.: One-Pot Synthesis of Fluorescent Carbon Graphene by the Exfoliation of Graphite in Ionic Liquids. ACS Nano. **200**, ACS Nano. **3**, 2367–2375 (2009)
35. Li, H., He, X., Kang, Z., Huang, H., Liu, Y., Liu, J., Lian, S., Tsang, C.H.A., Yang, X., Lee, S.T.: Water-soluble fluorescent carbon quantum dots and photocatalyst design. Angew. Chemie Int. Ed. **49**, 4430–4434 (2010). https://doi.org/10.1002/anie.200906154
36. Ye, R., Peng, Z., Metzger, A., Lin, J., Mann, J.A., Huang, K., Xiang, C., Fan, X., Samuel, E.L.G., Alemany, L.B., Martí, A.A., Tour, J.M.: Bandgap engineering of coal-derived graphene quantum dots. ACS Appl. Mater. Interfaces. **7**, 7041–7048 (2015). https://doi.org/10.1021/acsami.5b01419
37. Ding, H., Yu, S.B., Wei, J.S., Xiong, H.M.: Full-color light-emitting carbon dots with a surface-state-controlled luminescence mechanism. ACS Nano **10**, 484–491 (2016). https://doi.org/10.1021/acsnano.5b05406

Silica Nanoparticles: Morphology and Applications

Luiza Schwartz Dias and Annelise Kopp Alves

Abstract Morphology-controlled nanomaterials have been leading groundbreaking technological discoveries. The biocompatibility of silica nanoparticles presents themselves as a critical asset for their use in biomedicine as nanocarriers in drug delivery and bioimaging. Combining characteristics like high surface area, ease of surface functionalization, controllable pore size, and pore volume, nanosilica plays a crucial role in the future of materials science, manufacturing, environmental solutions, food, and energy. Based on the up-to-date literature review, a panorama about recent advances in synthesis and morphology aspects of nanosilica will be presented in this chapter and its application in food, energy, catalysts, and biomedicine.

Keywords Silica nanoparticles · Nanosilica · Drug delivery · Surface functionalization · Nanocarriers

Abbreviations

ACT	Adoptive T-Cell Immunotherapy
BMD	Bicontinuous microemulsion droplet
C' Dots	Cornell Dots
CdTe	Cadmium telluride
CTAB	Cetyl trimethylammoniun bromide
DFNS	Dendritic Fibrous Nanosilica
DSSC	Dye-sensitized solar cells
EU	European Union
FDA	United States Food and Drug Administration
GRAS	Generally Regarded Safe
HCl	Hydrochloric acid
HSC	High Strength Concrete
HSNP	Hollow Silica Nanoparticles

L. S. Dias (✉) · A. K. Alves
Universidade Federal do Rio Grande do Sul, Porto Alegre, Brazil

A. Kopp Alves (ed.), *Technological Applications of Nanomaterials*,
Engineering Materials, https://doi.org/10.1007/978-3-030-86901-4_5

IB	Ibuprofen
KAUST	King Abdullah University of Science and Technology
KCC-1	King Abdullah University of Science and Technology Catalysis Center
LWC	Light weight cement
MCM-41	Mobil Composition of Matter n. 41
MRI	Magnetic Resonance Imaging
MSNS	Mesoporous Nanosilica
NS	Nanosilica
NSHS	Nanosilica hollow spheres
O/W	Oil-in-water
PEG	Poly ethylene glycol
QD	Quantum Dots
RES	Reticuloendothelial system
SAS	Synthetic Amorphous Silica
SBA-15	Santa Barbara Amorphous Silica type 15
SDA	Structure-Directing Agents
SiNPs	Silica nanoparticles
SiO_2	Silicon Dioxide, Silica
SNC	Silica-based nanocapsules
TEOS	Tetraethylortosilicate
USA	United States of America
UV	Ultraviolet light

1 Introduction

Silica nanoparticles (SiNPs) are widely used in the materials science field as precursors to new technologies development. In recent years, the use of silica as a nanomaterial has attracted researchers' attention worldwide due to its exceptional properties such as low density, high surface area, and pore volume, good biodegradability, tunable pore shape, and size, alongside the ease of surface functionalization [1, 2]. Besides, thanks to its biocompatibility, silica is highly estimated and held as "generally regarded safe" (GRAS) by the United States Food and Drug Administration (FDA) [3]. Moreover, silicon (Si) sits as the second-most abundant chemical element in the Earth's crust [4], being found mostly in the form of quartz, crystalline silica.

SiNPs are extensively used as raw material in industries added to ceramics, electronic components, pharmaceutics, catalysts, etc. Thanks to the high surface area, ultra-fine amorphous silica powders brought forth many technological applications. In the form of silica gel (SiO_2 gel), it presents a tremendous ability to adsorb moisture relevant to the biomedical field. Furthermore, the highly porous structure makes it possible to use amorphous SiNPs as drug delivery agents by adsorbing large amounts of active components onto its surface [5]. Therefore, nanosilica (NS) garnered more attention from the biological and pharmaceutical sector over the last years due to its ability to carry therapeutic agents on cutting-edge technologies, such as encapsulation of drugs, fluorescent and magnetic nanoparticles for drug delivery and image sensing [6–9].

Engineered NS has been used in several industries to improve products quality. The construction sector enhances concretes properties by adding nanosilica, and the literature results showed increased compressive strength and durability, reduced pore-size distribution, and sorptivity [10–12]. SiNPs have been used in powdered food as anticaking and antifoaming agents, amongst other food industry applications [1]. For better ultraviolet (UV) protection, long-lasting benefits enhanced color, and finish quality, silica nanoparticles are added in the cosmetic formulation of several leave-in and wash-off products [13, 14]. The ever-growing application of SiNPs in the biomedical field is perceptible by the number of studies carried out over the past few years [5–9].

Many endeavors are made to develop new synthesis routes able to produce distinct morphologies. For instance, a noteworthy way to produce colloidal SiNPs is the sol–gel process based on the Stöber method [15], which is the cornerstone of many of the newest synthesis processes. One of the most recent researches in this segment focuses on developing dendritic fibrous nanosilica (DFNS) through emulsion systems [9, 16, 17]. To synthesize mesoporous silica nanoparticles (MSNs) with adjustable pore size, the route consists of a supramolecular assembled surfactant that serves as a template to conduct the framework [7].

Even though a global interest in the SiNPs performance exists, mainly for biomedical purposes, there is yet a lot to be discovered about its morphologies, applications, and synthesis process routes. Therefore, in this chapter, recent literature is reviewed to provide a panorama about advances in both synthesis and morphology aspects of SiNPs. Furthermore, the applications of such nanomaterial in biomedical research, energy storage, sensors, and the food industry are also outlined herein.

2 Silica Nanoparticles

Silica is a convenient abbreviation for silicon dioxide (SiO_2); it can be found in crystalline, amorphous, and moisturized form, or the hydroxylated form, also designated as silanol or siloxane [18]. In addition, this chemical compound is found in tetrahedral units (SiO_4), which allows forming a three-dimensional crystalline network through the oxygen bond with neighboring groups (Fig. 1) [19].

Silica nanomaterials first started gathering attention due to Stöber et al. pioneering process, capable of synthesizing monodisperse silica col-loids by using hydrolysis of alkyl silicates following condensation of silicic acid using ammonia as catalyst [15]. This discovery enabled further findings in the field of silica materials, such as mesoporous silica, namely Mobil Composition of Matter n. 41 (MCM-41) and Santa Barbara Amorphous Silica type 15 (SBA-15). For that matter, a significant interest arose in the design and synthesis of NS with several shapes, sizes, morphologies, and properties, such as tunable pore size and pore volume, for instance.

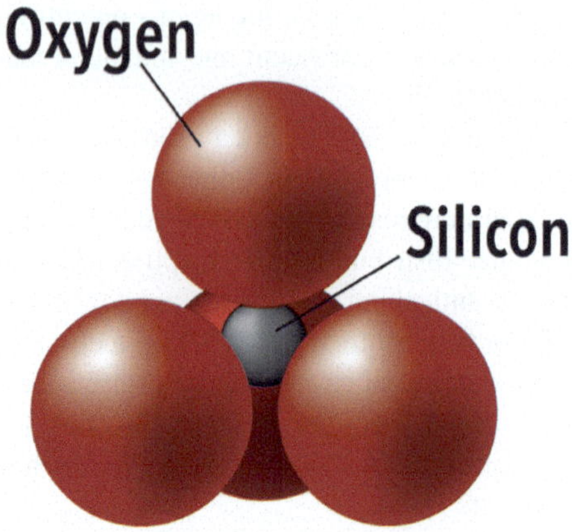

Fig. 1 Silica tetrahedral unit. *Source* Karla Panchuk (2018) CC BY-SA 4.0. Modified after Helgi (2013) CC BY-SA 3.0

Furthermore, most of the appeal to SiNPs is due to silanol groups in their surface, which allows a variety of chemical modifications. Indeed, silica surface modifications enable more versatile nanomaterials development, with specific properties related to the chemicals linked to it. The porous nature of silica, its tunable pore size, and distribution allows substances, such as drug molecules to be confined within its pores [6].

The SiNPs morphology will be determined mainly by the synthesis route chosen. For instance, it is commercially common to find silica gel synthesized through a sol–gel process using tetraethylorthosilicate (TEOS) or via quartz sand fusion mixed with sodium carbonate at high temperatures (>1300 °C) [20]. However, these methodologies have substantial costs engendered due to energy consumption and equipment investment [21].

Mainly, silicon occurs in the soil as soluble, adsorbed, or precipitate Si. The interactions between water and soil, silica-rich, are paramount to plants and other living beings [22]. Beyond nutritional, the Si absorbed by plants can strengthen them, avoiding plague attacks and disease spreading [23, 24]. Besides, a Si-rich precursor is paramount to SiNPs synthesis.

For this reason, recent studies show the search for new precursors as Si source, pursuing a less expensive synthesis process and more environment-friendly, such as rice husk ash [5, 24–26], pine cones and needles [27], pumice [20], fly ash [28], corn cob ash [29], wheat straw [30], among others.

3 Morphologies

Porous and highly dispersed components have earned space and interest from academic and industrial communities over the years, especially those with tunable pore diameters ranging from 2 to 50 nm [31, 32]. Amongst them, we find mesoporous

Table 1 A comparison of the physicochemical properties of DFNS, MCM-41, SBA-15, and MSN

Name	DFNS	MCM-41	SBA-15	MSN
Pore structure				
TEM	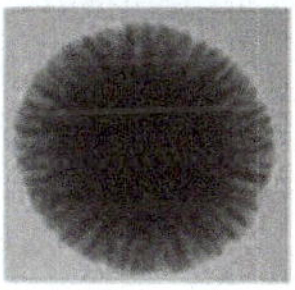	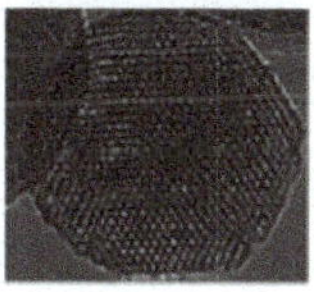		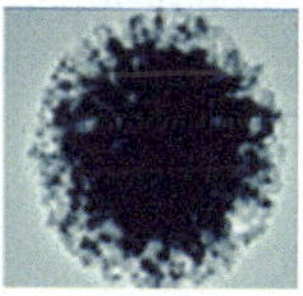
Pore sizes (nm)	3–25	30	5–50	2.88–9.92
Pore volume (cm^3 g^{-1})	0.54–2.18	0.7–1.2	0.75–1.15	0.85–0.95

Adapted from [16]—Published by The Royal Society of Chemistry

nanosilica to be one of the most versatile due to its high surface area and high structural order [7, 33]. Furthermore, MSNs presents several mesopores distributed in their honeycomb-like structure; as a consequence, they exhibit tremendous advantages, such as surface area greater than 1000 m^2g^{-1}, high pore volume (0.5–2.5 cm^3g^{-1}), adjustable pore size (1.3–30 nm), appreciable thermal and chemical stability, proper biocompatibility and biodegradability [34, 35].

A comparison between some morphology approached in this chapter is presented in Table 1.

3.1 *Mobil Composition of Matter N. 41 (MCM-41)*

In this area of research, the 1990s became memorable because of the groundbreaking synthesis of MSNs using surfactants as structure-directing agents (SDAs) [36]. Following Stöber's ingenious monodispersed silica synthesis method development, Kresge et al. presented a novel work on the field of mesoporous materials with the creation of MCM-41 [36]. This innovative protocol was developed through a liquid-crystal "templating" mechanism, creating ordered hexagonal mesopores.

The characteristics bound into MSNs structures allow them to encapsulate drugs (or bioactive agents) while protecting them from enzymatic degradation induced by the surrounding media [8]. Therefore, in early 2000, Vallet-Regi and his fellows introduced an MCM-41 novel application. That is to say, they encapsulated ibuprofen

(IB) into the nanomaterial and performed an in vitro drug release test under body fluid stimuli [37].

Conventionally, MCM-41 was synthesized using reactants hydrothermal processing in an autoclave; nowadays, it is widely processed through micro-wave irradiation because of the benefits of process control advancements [38]. The key parameter which determines pore shape is the surfactant used as a template. However, as well as the surfactant, the silicon source, the temperature, and the pH are essential parameters influencing MCM-41 structures [39]. The synthesis procedure is a modification of the method described by Stöber in 1968. In both methods, hydrothermal and microwave irradiation, a homogeneous solution is formed by mixing a template and the silicon source. After adjusting the pH, the solution is heated through one of the mentioned methods to synthesize MCM-41 [40]. The MSNs are achieved using a sol-gel process catalyzed in an alkaline media [41]. To achieve better crystallinity and remaining surfactant removal, a step of calcination of the synthesized powder is performed in the end.

3.2 *Santa Barbara Amorphous Silica Type 15 (SBA-15)*

The synthesis process of MSNs production has been profoundly enhanced by using polymeric templates in its preparation over the years. In 1998, Zhao and coworkers implemented an organic structure-directing agent, the amphiphilic triblock copolymer, to obtain hexagonal mesoporous silica structures, which resulted in uniform pore sizes as big as approximately 300 angstroms [42]. The nanomaterial created was named Santa Barbara Amorphous type 15 (SBA-15) due to its origins in the University of California laboratories in Santa Barbara (USA). Notably, in early 2000 the hexagonal phase SBA-15 was the most outstanding member of the block copolymer templated materials family due to its uniqueness in synthesis conditions, tailored particle morphology, pore size tunability, and hydrothermal stability [43]. Comparing SBA-15 to MCM-41, the former presents larger pore sizes and thicker walls than those of the latter, which aids faster drug release [44]. This mesoporous nanosilica sieve displays properties such as high surface area, chemical inertness, well-defined porous architecture, robust physico-chemical properties, thermal and mechanical stability, and sufficient active sites for grafting several functional chemical groups [45]. Hence, SBA-15-based composites, with the incorporation of active components, have been attracting growing attention in several research fields. Surface modification processes via post-synthesis grafting (silylation) with alkoxysilanes are popular methods to incorporate functional groups into the materials sieves [46].

SBA-15 synthesis typically consists of a template (amphiphilic triblock copolymer—$EO_{20}PO_{70}EO_{20}$, Pluronic P123) dissolved in acidic solution (water and HCl mix) by the slow silica source addition under continuous stirring at moderate temperature. To achieve silica gel precipitation, the synthesis requires strong acidic conditions (pH approximately 1). Usually, the silica source used is TEOS, which is dropwise added and stirred into the solution. Subsequently, the suspension is aged at an elevated temperature overnight and then a white solid should be retrieved through filtration, washing, and drying for over 12 h [42]. As well as made in the MCM-41, a final calcination in air is made to obtain SBA-15 free of the surfactant template [47].

3.3 Nanocapsules

The growing interest in new ways of encapsulating bioactive molecules for controlled drug delivery or bioimaging applications made silica-based nanocapsules (SNCs) an attractive asset, especially in biomedical applications. Such nanomaterial combines the benefits of silica, namely biocompatibility and versatility in surface functionalization, with capsular configurations, such as a broad inner cavity, low density, and good colloidal properties [48].

SNCs are structures composed of a solid shell that encloses a core-forming space used to encapsulate active molecules for several purposes [6]. Although the inner cavity might be used to contain specific molecules, it can also confine air. This specific balloon-like structure, unique in shape configuration, is named hollow silica nanoparticles (HSNPs) [49, 50].

The ease of surface modification of the silica shell via functional moieties (such as polymers and antibodies) through siloxane chemistry allowed SNCs to gain multifunctionality. To build up nanosilica hollow spheres (NSHS) performance, a lot of effort has been made in the silica shell interfacial engineering, similarly in silica framework (siloxane bonds), as well as in the surface (silanol groups) level.

Since the silica capsules synthesis templating precursor work [51, 52], several methodologies used to prepare various silica capsules types were developed.

Usually, HSNPs synthesis routes can be split into two main methods: template and non-template [49]. Using the non-templated method, one can follow the electrostatic atomization method or the spray pyrolysis method, which allows mass production of the HSNPs, characterized by its polydispersity.

To improve shape control of the particles, it is commonly used a template method that can be summed up in three main steps. The first step consists of template preparation. Over the past years, sacrificial templating has been broadly used in silica hollow nanoparticles/ nanocapsules synthesis [53]. Emulsions, considered soft templates, are often used in situations of reduced synthesis complexity. They comprehend a mixture of two immiscible liquids, with one dispersed over the other in a way to form an oil-in-water (O/W) emulsion, which can be loaded with active materials in advance [54]. The final capsule size is substantially determined by the emulsion droplet [55]. The second step is forming the shell, silica-based in this case, onto the template surface. Therefore, shell materials, as silica, are deposited over the O/W interface of an emulsion template, containing all sorts of active agents encapsulated inside the core. Shell formation mechanism might be slightly different according to the catalyst used, which can also influence the reaction rate and final product. The third and final step is template removal, and it might be done either through calcination or using a proper solvent to dissolve the template and form the inner shell cavity.

3.4 Dendritic Fibrous Nanoparticles (DFNS)

The successor of MSNs, dendritic fibrous nanosilica (DFNS), arose as an important development amongst morphology-controlled nanomaterials. It was first presented in 2010 by Polshettiwar and coworkers at KAUST (King Abdullah University of Science and Technology) Catalysis Center; therefore, it is also called KCC-1 [16]. These nanomaterials are some sort of three-dimension (3D) fibrous spheres presenting center-radial channels with precise mesostructures developed from cyclohexane emulsion systems. Despite being named "fibrous", DFNS fibers are closer to thin sheet shape, varying from 3.5 nm to 5.2 nm thickness, analogous to flower petals. Therefore, it can be found in the literature under other names, such as wrinkled, dendritic, dandelion, or lamellar [56, 57].

The dendritic nanosilica is suitable for high pressure and high-temperature applications; hence its excellent thermal (up to 800 ºC), mechanical (up to 216 MPa), and hydrothermal stability [57, 58]. Not to mention, DFNS has a distinct characteristic of tunable pore size and pore volume, especially when it is compared to other MSNs. The accessibility improved by its form allowed increased active sites loading over the nanosilica surface. Furthermore, the tailorable pore size and pore volume can be made for specific guest molecules with different sizes and loadings spaces [56]. DFNS pores can be filled with organometallic complexes, metals, peptides, enzymes, proteins, carbon, and so forth. Equally important is the accessibility of its pores from all sides (not possible in MCM-41 and SBA-15 tubular pores), also decreasing pore (channel) blockage occurrence. However, dendritic fibrous nanospheres' surface area can be lower than those found in either MCM-41 or SBA-15 [57]. DFNS are also considered disordered when compared to ordered mesoporous silica-based nanomaterials, such as those of MCM e SBA families [59].

The surface area of these fibrous nanospheres may vary from 450 up to 1244 m^2g^{-1} using particle size and fiber density adaptability [61]. Unlike traditional silica channel pore size distribution, DFNS has radially oriented pores (also known as V-shaped) that increase its size from the sphere center towards its edges. The initial adsorption of guest molecules is profoundly helped by the macropores present on the nanomaterial studied.

Originally, Polshetiwar et al. [58] synthesized KCC-1 through a micro-wave-assisted hydrothermal technique. First of all, the authors homogeneously mixed cetyltrimethylammonium bromide (CTAB), water, and urea [60]. Afterward, it was added, dropwise, the solution of TEOS and p-Xylene (or cyclohexane). An emulsion was the expected result after mixing the two prior solutions; this is where the lamellar phase of CTAB occurs. At this point, with a homogeneous mix of the two solutions, a co-surfactant is added to stabilize even more the emulsion, forming the bicontinuous microemulsion droplet (BMD), which plays a role of a template for silica growth. At this stage, there are two channels (water and oil) in the lamellar phase. Throughout these BMDs, the silica precursor (TEOS) diffuses to the oil–water interface, becoming hydrolyzed. TEOS hydrolysis is aided by the urea decomposition when it forms ammonium hydroxide during the reflux step through regulating the reaction pH. Supersaturation is triggered when a certain saturation of hydrolyzed TEOS molecules is reached in the water channel, all followed by a growth step. The BMDs and the reaction time are paramount factors that control the size, shape, and morphology of the nanosilica. The reaction mixture is kept in reflux for several hours, where water channels are then filled by silica while the oil fraction remains empty. The water channels thickness will guide the resulting fiber thickness (silica sheet), while the oil channels width dictates the pore size. To obtain pure DFNS, the solid achieved must be isolated, washed, and calcined for 6 h at 550 °C in air [16].

4 Applications

4.1 Food

Nanosized particles are predicted to attract even more attention in food production, processing, packaging, and/or storage in the future to come [61]. Amongst those nanomaterials, synthetic amorphous silica (SAS) represents a very routine food addictive over the past decades [62]. That is to say, SAS has been used as a food additive for about 50 years now. Composed of Si aggregates of primary particles, SAS displays sizes in the lower nanometer range [63].

In the United States of America (USA), the FDA approves the use of silica particles only in amounts smaller than 2% by weight [64]. Furthermore, the Commission Regulation 1129/2011 of the European Union (EU) makes more rigid control over SiNPs presence and allows its level only up to 1% by weight (in drier powdered foodstuffs) [65]. Not to mention, the EU only allows the usage of SiNPs as food additives if they were synthesized by thermal (pyrogenic) or wet (precipitation) processes [66].

SAS is extensively used as an additive in processed food under the code E551, registered by the EU. Usually, its main purpose in the food industry is to avoid poor flow ("caking"), but it is also employed as a carrier of flavor/aroma, to clarify/fining beverages, as antifoaming [67].

MSNs are also found in the food industry as catalysts in several processes, such as emulsification and packaging. Table 2 displays some of the studied particles and their applications.

HMS, hexagonal mesoporous silica; MCM-41, mobile crystalline material No. 41; MSU-X, Michigan State University; RL-SBA-15, rod-like SBA-15; SBA 15, Santa Barbara amorphous-type material. Reproduced with permission from [3] ©2013, Czech Journal of Food Sciences.

Table 2 Catalytic applications of mesoporous silica nanoparticles in the food industry

Mesoporous support	Catalyst	Molecule synthesized	Application in food
MCM-41	Sulfonic acids	Fatty acids	Emulsifier
MCM-41	Pd_6Ru_6/SiO_2	Nylon-6	Food packaging
MCM-41	Ru_6Sn/SiO_2	Nylon-6	Food packaging
HMS	Heteropolyacid	6-Allyl-2,4-di-tert-butylphenol	Valuable products
Mesostructured solid	Sulfonic acid	Fatty acid methyl ester	Biodiesel production
SBA—15	Enzymes	Biomolecules	Food research
RL-SBA-15	Enzymes	Biomolecules	Food research
MSU-X	Enzymes	Biomolecules	Food research
Nanoporous silica glass	Enzymes	Sugars	Food industry
Nanoscale carrier	Enzymes	Biofilm removal	Food processing industry
Epoxy sílica nanoparticles	Xylitol dehydrogenase	l-Xylulose	Food industry
Silica nanostructured	Au	Lactobionic acid	Food industry

4.2 Building Materials

Nanoparticles can be found all over amongst the construction industry materials. However, one of the most significant research areas is the addition of nanosilica in cement for concrete enhancement, being the object of recent studies that demonstrated the advantages of its usage [12, 68]. The addition of NS into cement paste showed positive results, such as less cement consumption in high strength concrete (HSC) production, lower carbon dioxide (CO_2) footprint related to the process, and higher cost-effective cement products [69].

NS also presents effects on concrete properties compared to other clinkers due to its superior reactivity, high surface ratio, and nanosized particles [10]. The SiNP addiction into cement-based products may affect its properties in three main ways; the first is in the nucleation, where it accelerates the cement hydration; second is in the filling, causing it to strengthen the material microstructure; and finally, the third is related to the pozzolanic reaction, producing supplementary C-S-H gel, enhancing concrete mechanical properties [70–72].

Research shows that the incorporation of silica nanoparticles into lightweight cement (LWC) can improve its compressive strength (after 1 and 7 curing days), decrease water sorptivity and its penetration depth, together with a drop in the chloride ion permeability [11].

4.3 Energy

In the search for more efficient and cost-effective solar energy generators, hybrid solar converters gained the attention of researchers [73]. One goal is to utilize converters that can store as heat the excess energy that cannot be directly used to produce electricity [74]. Therefore, a material capable of efficiently converting light into heat is required. One promising new approach is the mesoporous silica capsules embedding immobilized plasmonic gold (Au) nanoparticles utilization as black metal alternatives. The plasmon band broadness can be ensured by the silica shell effectiveness of Au particles immobilization in its inner interface [73].

Another energy-related application is the utilization of DFNS on light-harvesting in solar cells. The use of TiO_2-coated fibrous nanosilica as a scatterer in dye-sensitized solar cells (DSSCs) is the focus of studies, thus presenting a correlation between nanospheres morphology and photovoltaic performance [75]. DFNS may also be used in photocatalytic organic transformations, besides its applications in water splitting, dye degradation, and solar cells [57].

4.4 Catalysts

Over the recent years, mesoporous molecular sieves (pore diameter range 2–50 nm) have attracted research and industry attention in the catalysis sector [60]. The application of such nanomaterial ranges from catalyst supports to adsorbents. For that matter, mesoporous nanosilica presents itself as a good fit due to its high surface area, tunable pore size, and volume, apart from its well-defined pore structure [76].

One vital characteristic of MSN (i.e., MCM-41) for applications in catalysis is the presence of two distinct surfaces, the internal one (inside the pores) and one which is external (nanoparticle surrounding interface). Therefore, the ability to modify MSNs surfaces with functional groups, through functional groups, makes them overly versatile. For instance, two main methodologies allow MCM-41 functionalization, namely, grafting and co-condensation [41]. While grafting might be a more straightforward way to achieve surface functionalization (single step), usually, co-condensation is more practical when a homogeneous and uniform surface is desired. Moreover, Martínez-Edo and coworkers found that MCM-41 functionalization may allow the preparation of a more advanced system incorporating metallic complexes, catalyzing palladium cross-coupling reactions, cycloadditions, and hydroformylations [41].

On the other hand, nanoparticles like dendritic fibrous nanosilica present V-shaped pores, thus improving pore accessibility and allowing reactant molecules diffusion. Initial guest molecule adsorption may also benefit from DFNS macropores. They also will enable a series of metal nanoparticles to be loaded onto its pores, thus aiding catalytic transformations [77].

4.5 Biomedical/ Pharmaceutical

4.5.1 Drug Delivery

SiNPs have been researched as drug delivery vials for improved solubility (biological, chemical, and physical), controlled, and targeted drug release. In addition, engineered nanosilica is also used by nanomedicine in cancer targeted delivery since its ability to encapsulate specific nanocarriers, such as liposomes, micelles, and drug conjugates.

MSNs may provide a brand new arsenal of functionalities not covered by conventional medicine, like drug specificity, broader efficacy window of some medicines, both drug solubility and/or stability improved, less detrimental pharmacokinetic profiles, and fewer side effects [32].

In their study, Thiramanas and coworkers [54] found nanocapsules made of silica to be suitable for uptake in difficult to be transfected T-Cells. The ease of structure and surface manipulation brings SiNC into the spotlight of a promising delivery system that can be used as a nanocarrier in adoptive T-Cell immunotherapy (ACT). This kind of treatment uses immune cells from patients themselves to destroy cancer cells and is considered a breakthrough in cancer therapy [78]. In their study, one-pot synthesis was used to incorporate the therapeutic payload into the inner cavity of a silica shell nanocapsule, allowing high loading capacity and cargo protection [54].

The surface functionalization (inner pore system and/or external particle surface) are the key that allows MSNs to become target-specific [2]. Furthermore, the combination of MSNs with numerous polymers has widely expanded its applications in controlled drug delivery systems. The utilization of polymers as gatekeepers or capping agents may aid drug release control, enable sustained drug release, lower adverse side effects and boost the encapsulated drugs therapeutic index.

For instance, nanoparticles polyethylene glycol (PEG) coated present themselves as a suitable approach to prevent either reticuloendothelial system (RES) nonspecific uptake and biomacromolecules nonspecific interaction, besides it also aids in water solubility improvement [79, 80].

Similarly, the confinement of water-insoluble drug molecules within MSNs may reduce amorphous drug crystallization, thus reducing lattice energy and improving both dissolution rate and bioavailability [81–83]. Not to mention that MSNs vast hydrophilic surface also promotes wetting and stored drug dispersion, thus boosting its dissolution [84].

4.5.2 Bioimaging

High-end detection systems using nanoparticles might be the future of noninvasive diagnosis approaches [1]. For instance, in bioimaging applications, a variety of fluorescent probes types and magnetic nanocrystals may be incorporated onto silica-based nanocapsules due to the latter high colloidal and chemical stability in physiological media photosensitive compounds protection while maintaining proper image contrast [48]. For this reason, SNCs imaging capabilities have been used for fluorescence imaging, Magnetic Resonance Imaging (MRI), or a combination of these modalities.

To exemplify nanosilica potential in the biomedical field, in 2011 FDA approved the Si-based drug called Cornell Dots (C' dots) to be used in cancer molecular imaging for first-in-human clinical trials [85]. C' dots were built to locate cancer cells inside the body using core–shell nanoparticles mostly made of silica.

Lowe et al. performed a test in tumor-bearing mice where PEGylated Fe_2O_4/dye@SNCs were injected to observe its efficiency [33]. The MRI showed that it was possible to observe a significant amount of SNCs were detected in the tumor area after 40 min, which was confirmed afterward with further testing. When conjugated with antibodies or peptides, SiNPs can selectively target cancer cells, thus displaying cancer-specific and differential color imaging at single wavelength excitation, either in vitro and in vivo [86].

Developed for Zhang et al., another approach for bioimaging is the use of DFNS to stabilize cadmium telluride (CdTe) Quantum Dots (QD) [87]. The novel approach displayed better QD stability, improved cytotoxicity, and strong fluorescence, thus important for bioimaging [58]. For applications such as specific immune cells identification, Walczyk and coworkers developed fluorescent dye-conjugated MSNs with promising results [88].

5 Obstacles and Future Perspectives

All things considered, SiNPs have risen as a key innovation in nanomaterial science over the last years, attracting a great deal of attention in fields like energy, environment, and health. The notable features of such nanomaterials include high surface area, tunable pore size, exceptional biocompatibility, and great potential for encapsulating various molecule types. Furthermore, depending upon preparation strategy, silica-based nanomaterials can present different morphologies, such as MCM-41, SBA-15, hollow spheres, nanocapsules, dendritic fibrous, and so on.

The ease of surface functionalization combined with the range of possible morphologies makes SiNPs play a promising role in bioimaging, drug delivery, and cancer therapy.

Although nanosilica has been widely researched in the biomedical area, the critical challenge still remains in the silica pharmacokinetic and pharmacodynamics investigation in vivo diagnostic and therapeutic applications.

Overall, SiNPs have presented exciting progress over the past few decades in synthesis, surface modification, thus confirming their relevance in future biomedical applications as multifunctional carriers, amongst other fields.

References

1. Mebert, A.M., et al.: Nanoengineered silica: Properties, applications and toxicity. Food Chem. Toxicol. **109**, part 1 (2017). https://doi.org/10.1016/j.fct.2017.05.054
2. Singh, P., et al.: Nanosilica: recent progress in synthesis, functionalization, biocompatibility, and biomedical applications. ACS Biomater. Sci. Eng. **5**(10), 4882–4898 (2019). https://doi.org/10.1021/acsbiomaterials.9b00464
3. Bernardos, A., et. al.: Applications of mesoporous silica materials in food—a review **31**(2), 99–107 (2013). https://doi.org/10.17221/240/2012-CJFS
4. Lide, D.R.: Abundance of Elements in the Earth's Crust and in the Sea, in CRC Handbook of Chemistry and Physics, p. 2373. Ed. Boca Raton, CRC Press LLC, FL (2005)
5. Pijarn, N., et al.: Synthesis and characterization of nanosized-silica gels formed under controlled conditions. Powder Technol. **203**(3), 462–468 (2010). https://doi.org/10.1016/j.powtec.2010.06.007
6. Wibowo, D., et al.: Interfacial engineering for silica nanocapsules. Adv. Colloid Interface Sci. **236**, 83–100 (2016). https://doi.org/10.1016/j.cis.2016.08.001
7. Farjadian, F. et. al.: Mesoporous silica nanoparticles: Synthesis, pharmaceutical applications, biodistribution, and biosafety assessment. Chem. Eng. J. **359**(June 2018), 684–705 (2019). https://doi.org/10.1016/j.cej.2018.11.156
8. Bao, B.Q., et al.: Evolution and present scenario of multifunctionalized mesoporous nanosilica platform: A mini review. Mater. Sci. Eng. C **91**(June), 912–928 (2018). https://doi.org/10.1016/j.msec.2018.07.008
9. Maity, A., et al.: Dendritic fibrous nanosilica (DFNS) for RNA extraction from cells. Langmuir **36**(42), 12755–12759 (2020). https://doi.org/10.1021/acs.langmuir.0c02520
10. Elrahman, M.A. et. al.: Influence of nanosilica on mechanical properties, sorptivity, and microstructure of lightweight concrete. Materials (Basel). **12**(19), (2019). https://doi.org/10.3390/ma12193078
11. Du, H.: Properties of ultra-lightweight cement composites with nano-silica. Constr. Build. Mater. **199**, 696–704 (2019). https://doi.org/10.1016/j.conbuildmat.2018.11.225
12. Du, H., et al.: High performance cement composites with colloidal nano-silica. Constr. Build. Mater. **224**, 317–325 (2019). https://doi.org/10.1016/j.conbuildmat.2019.07.045
13. Zaccariello, G., et al.: Formation and controlled growth of bismuth titanate phases into mesoporous silica nanoparticles: An efficient self-sealing nanosystem for UV filtering in cosmetic formulation. ACS Appl. Mater. Interfaces **9**(2), 1913–1921 (2017). https://doi.org/10.1021/acsami.6b13252
14. Tolbert, S.H., et al.: New hybrid organic/inorganic polysilsesquioxane-silica particles as sunscreens. ACS Appl. Mater. Interfaces **8**(5), 3160–3174 (2016). https://doi.org/10.1021/acsami.5b10472
15. Stöber, W., et al.: Controlled Growth of monodisperse silica spheres in the micron size range. J. Colloid Interface Sci. **26**, 62–69 (1968). https://doi.org/10.1016/0021-9797(68)90272-5
16. Shaban, M., et al.: Biomedical applications of dendritic fibrous nanosilica (DFNS): Recent progress and challenges. RSC Adv. **10**(61), 37116–37133 (2020). https://doi.org/10.1039/D0RA04388E
17. Thankamony, A.S.L., et al.: Insights into the catalytic activity of nitridated fibrous silica (KCC-1) Nanocatalysts from ^{15}N and ^{29}Si NMR Spectroscopy Enhanced by Dynamic Nuclear Polarization. Angew. Chemie Int. Ed. **54**(7), 2190–2193 (2015). https://doi.org/10.1002/anie.201406463
18. Jadhav, SA. et al.: Recent advancements in silica nanoparticles based technologies for removal of dyes from water. Colloids Interface Sci. Commun. **30**(April), 100181 (2019). https://doi.org/10.1016/j.colcom.2019.100181
19. Bergna, H.E. et. al.: Colloidal silica: Fundamentals and applications. Colloid. Silica Fundam. Appl. 1–897 (2005). ISBN: 978–1–4200–2870–6

20. Mourhly, A. et. al.: Highly efficient production of mesoporous nano-silica from unconventional resource : Process optimization using a central composite design. Microchem. J. **145**(October 2018), 139–145 (2019). https://doi.org/10.1016/j.microc.2018.10.030
21. Gomes, L.S., et al.: About silica and its particularities. Rev. Virtual Química **10**(4), 22 (2018). https://doi.org/10.21577/1984-6835.20180072
22. Karimi, J., et al.: Effects of silicon oxide nanoparticles on growth and physiology of wheat seedlings. Russ. J. Plant Physiol. **63**(1), 119–123 (2016). https://doi.org/10.1134/S1021443716010106
23. El-Shetehy, M. et al.: Silica nanoparticles enhance disease resistance in Arabidopsis plants. Nat. Nanotechnol. **16**(3), 344–353 (2021). https://doi.org/10.5281/zenodo.4131137
24. Zarei, V.: Environmental method for synthesizing amorphous silica oxide nanoparticles from a natural material. Processes **9**(2), 1–9 (2021). https://doi.org/10.3390/pr9020334
25. Srivastava, K. et. al.: A facile method for production of amorphous silica from perlite under microwave irradiation. Int. J. IT, Eng. Appl. Sci. Res. **4**(1), 2319–4413 (2015). ISSN: 2319–4413
26. Mor, S., et al.: Nanosilica extraction from processed agricultural residue using green technology. J. Clean. Prod. **143**, 1284–1290 (2017). https://doi.org/10.1016/j.jclepro.2016.11.142
27. Assefi, M., et al.: Green synthesis of nanosilica by thermal decomposition of pine cones and pine needles. Adv. Powder Technol. **26**(6), 1583–1589 (2015). https://doi.org/10.1016/j.apt.2015.09.004
28. Yadav, V.K., et al.: A novel and efficient method for the synthesis of amorphous nanosilica from fly ash tiles. Mater. Today Proc. **26**, 701–705 (2019). https://doi.org/10.1016/j.matpr.2020.01.013
29. Okoronkwo, EA. et. al.: Extraction and characterization of amorphous silica from corn cob ash by sol-gel method. Chem. Mater. Res. **3**(4), 2225–956 (2013). ISSN: 2225–0956
30. Patel, K.G., et al.: Preparation and characterization of silica gel from wheat straw. Int. J. Chem. Eng. Appl. **7**(5), 344–347 (2016). https://doi.org/10.18178/ijcea.2016.7.5.603
31. Liu, P.P. et. al.: General Introduction to Porous Materials, in Porous Materials: Processing and Applications, 1st edn., pp. 1–20. Elsevier Inc. (2014). ISBN: 978–0–12–407788–1
32. Manzano, M., et al.: Mesoporous Silica Nanoparticles for Drug Delivery. Adv. Funct. Mater. **30**(2), 3–5 (2020). https://doi.org/10.1002/adfm.201902634
33. Lowe, J.B. et. al.: Deformation of Ordered Mesoporous Silica Structures on Exposure to High Temperatures, vol. 2014 (2014). https://doi.org/10.1155/2014/754076
34. Song, Y., et al.: Mesoporous silica nanoparticles for stimuli- responsive controlled drug delivery: advances, challenges, and outlook. Int. J. Nanomedicine **12**, 87–110 (2017). https://doi.org/10.2147/IJN.S117495
35. Mehmood, A., et al.: Mesoporous silica nanoparticles: A review. J. Develop. Drugs **6**, 174 (2017). https://doi.org/10.4172/2329-6631.1000174
36. Kresge, C.T., et al.: Ordered mesoporous molecular sieves synthesized by a liquid-crystal template mechanism. Nature **359**, 710–712 (1992). https://doi.org/10.1038/359710a0
37. Vallet-Regí, M., et al.: A new property of MCM-41: drug delivery system. Chem. Mater. **13**(11), 308–311 (2001). https://doi.org/10.1021/cm0011559
38. Dündar-Tekkaya, E. et. al.: Mesoporous MCM-41 material for hydrogen storage : A short review. Int. J. Hydrogen Energy **41**, 9789–9795 (2016). https://doi.org/10.1016/j.ijhydene.2016.03.050
39. Beck, J.S., et al.: A new family of mesoporous molecular sieves prepared with liquid crystal templates. J. Am. Chem. Soc. **114**, 10834–10843 (1992). https://doi.org/10.1021/ja00053a020
40. Dündar-Tekkaya, E. et. al.: Effect of loading bimetallic mixture of Ni and Pd on hydrogen storage capacity of MCM-41. Int. J. Hydrogen Energy, **40**, 7636–7643 (2015). https://doi.org/10.1016/j.ijhydene.2015.02.108
41. Martínez-Edo, G., et al.: Functionalized ordered mesoporous silicas (MCM-41): synthesis and applications in catalysis. Catalysts **8**(617), 1–62 (2018). https://doi.org/10.3390/catal8120617
42. Zhao, D. et al.: Triblock copolymer syntheses of mesoporous silica with periodic 50 to 300 Angstrom Pores **279**(January) (1998). https://doi.org/10.1126/science.279.5350.548

43. Hartmann, M. et. al.: Mechanical stability and porosity analysis of large-pore SBA-15 mesoporous molecular sieves by mercury porosimetry and organics adsorption. 18, 8010–8016 (2002). https://doi.org/10.1021/la025782j
44. McCarthy, C.A. et al.: Mesoporous silica formulation strategies for drug dissolution enhancement: A review Mesoporous silica formulation strategies for drug dissolution enhancement: a review. **5247**(November) 2015 (2016). https://doi.org/10.1517/17425247.2016.1100165
45. Yuan, S. et. al.: Recent advances of SBA-15-based composites as the heterogeneous catalysts in water decontamination: A mini-review. J. Environ. Manage. **254**(November 2019), 109787 (2020). https://doi.org/10.1016/j.jenvman.2019.109787
46. Pirez, C., et al.: Sol-gel synthesis of SBA-15: Impact of HCl on surface chemistry. Microporous Mesoporous Mater. **271**(May), 196–202 (2018). https://doi.org/10.1016/j.micromeso.2018.05.043
47. Verma, P. et al.: Functionalized mesoporous SBA-15 silica: Recent trends and catalytic applications. **12**(21). Royal Society of Chemistry (2020). https://doi.org/10.1039/D0NR00732C
48. Zhang, Y., et al.: Silica-based nanocapsules: Synthesis, structure control and biomedical applications. Chem. Soc. Rev. **44**(1), 315–335 (2015). https://doi.org/10.1039/C4CS00199K
49. Nakashima, Y., et al.: Synthesis of ultra-small hollow silica nanoparticles using the prepared amorphous calcium carbonate in one-pot process. Adv. Powder Technol. **29**(4), 904–908 (2018). https://doi.org/10.1016/j.apt.2018.01.006
50. Takai-Yamashita C. et. al.: Hollow silica nanoparticles: A tiny pore with big dreams. Adv. Powder Technol. **31**(2), 804–807 (2020). https://doi.org/10.1016/j.apt.2019.11.034
51. Schacht, S., et al.: Oil-water interface templating of mesoporous macroscale structures. Science **273**(5276), 768–771 (1996). https://doi.org/10.1126/science.273.5276.768
52. Caruso, F., et al.: Nanoengineering of inorganic and hybrid hollow spheres by colloidal templating. Science **282**(5391), 1111–1114 (1998). https://doi.org/10.1126/science.282.5391.1111
53. Abdelaal, H.M. et. al.: Microwave-based fast synthesis of clear-cut hollow spheres with mesoporous wall of silica nanoparticles as excellent drug delivery vehicles. J. Nanoparticle Res. **22**(7) (2020). https://doi.org/10.1007/s11051-020-04918-3
54. Thiramanas, R., et al.: Silica nanocapsules with different sizes and physicochemical properties as suitable nanocarriers for uptake in t-cells. Int. J. Nanomedicine **15**, 6069–6084 (2020). https://doi.org/10.2147/IJN.S246322
55. Graham, M., et al.: Formation Mechanism of Multipurpose Silica Nanocapsules. Langmuir **37**(2), 918–927 (2021). https://doi.org/10.1021/acs.langmuir.0c03286
56. Khantan, N., et al.: Synthesize of dendritic fibrous nano-silica functionalized by cysteine and its application as advanced adsorbent. Nanocomposites **5**(4), 104–113 (2019). https://doi.org/10.1080/20550324.2019.1669925
57. Maity, A., et al.: Dendritic Fibrous Nanosilica for Catalysis, Energy Harvesting, Carbon Dioxide Mitigation, Drug Delivery, and Sensing. Chemsuschem **10**(20), 3866–3913 (2017). https://doi.org/10.1002/cssc.201701076
58. Polshettiwar, V., et al.: High-surface-area silica nanospheres (KCC-1) with a fibrous morphology. Angew. Chemie - Int. Ed. **49**(50), 9652–9656 (2010). https://doi.org/10.1002/anie.201003451
59. Wang, Y., et al.: Dendritic fibrous nano-particles (DFNPs): Rising stars of mesoporous materials. J. Mater. Chem. A **7**(10), 5111–5152 (2019). https://doi.org/10.1039/C8TA09815H
60. Maity, A., et al.: Facile synthesis to tune size, textural properties and fiber density of dendritic fibrous nanosilica for applications in catalysis and CO2 capture. Nat. Protoc. **14**(July), 2177–2204 (2019). https://doi.org/10.1038/s41596-019-0177-z
61. Winkler, H.C., et al.: Critical review of the safety assessment of nanostructured silica additives in food. J. Nanobiotechnology **14**(1), 1–9 (2016). https://doi.org/10.1186/s12951-016-0189-6
62. Peters, R.J.B., et al.: Nanomaterials for products and application in agriculture, feed and food. Trends Food Sci. Technol. **54**, 155–164 (2016). https://doi.org/10.1016/j.tifs.2016.06.008

63. Van der Zande, M., et al.: Sub-chronic toxicity study in rats orally exposed to nanostructured silica,". Part. Fibre Toxicol. **11**(1), 1–19 (2014). https://doi.org/10.1186/1743-8977-11-8
64. US FDA 2015. United States Food and Drug Administration. Code of Federal Regulations Title 21, 21CFR172.480 (2015)
65. European Union, Commission Regulation (EU) No 1129/2011. Off. J. Eur. Union 1–212 (2011)
66. European Union, Commission Regulation (EU) No 231/2012. Off. J. Eur. Union 1–295 (2012)
67. Barahona, F., et al.: Multimethod approach for the detection and characterisation of food-grade synthetic amorphous silica nanoparticles. J. Chromatogr. A **1432**, 92–100 (2016). https://doi.org/10.1016/j.chroma.2015.12.058
68. Raheem, A.A., et al.: Incorporation of metakaolin and nanosilica in blended cement mortar and concrete- A review. J. Clean. Prod. **290**, 125852 (2021). https://doi.org/10.1016/j.jclepro.2021.125852
69. Abhilash, P.P., et al.: Effect of nano-silica in concrete: a review. Constr. Build. Mater. **278**, 122347 (2021). https://doi.org/10.1016/j.conbuildmat.2021.122347
70. Silvestre, J., et al.: Review on concrete nanotechnology. Eur. J. Environ. Civ. Eng. **20**(4), 455–485 (2016). https://doi.org/10.1080/19648189.2015.1042070
71. Shah, S.P., et al.: Nano-modification of cementitious material: Toward a stronger and durable concrete. J. Sustain. Cem. Mater. **5**(1), 1–22 (2015). https://doi.org/10.1080/21650373.2015.1086286
72. Aggarwal, P. et. al.: Use of nano-silica in cement based materials: a review. Cogent Eng. **2**(1) (2015). https://doi.org/10.1080/23311916.2015.1078018
73. ElKabbash, M., et al.: Tunable black gold: controlling the near-field coupling of immobilized Au nanoparticles embedded in mesoporous silica capsules. Adv. Opt. Mater. **5**(21), 1–6 (2017). https://doi.org/10.1002/adom.201700617
74. Xiong, Y., et al.: Enhanced thermal energy storage of nitrate salts by silica nanoparticles for concentrating solar power. Int. J. Energy Res. **45**(4), 5248–5262 (2021). https://doi.org/10.1002/er.6142
75. Kang, J.S. et. al.: Wrinkled silica/titania nanoparticles with tunable interwrinkle distances for efficient utilization of photons in dye-sensitized solar cells. Nat. Publ. Gr., (July), 1–14 (2016). https://doi.org/10.1038/srep30829
76. Chong, C.C., et al.: Development of nanosilica-based catalyst for syngas production via CO_2 reforming of CH4: A review. Int. J. Hydrogen Energy (2020). https://doi.org/10.1016/j.ijhydene.2020.01.086
77. Huang, Y.: Functionalization of mesoporous silica nanoparticles and their applications in organo-, metallic and organometallic catalysis, 10974 (2009). https://lib.dr.iastate.edu/etd/10974
78. Cohen, J.E., et al.: Adoptive cell therapy: Past, present and future. Immunotherapy **9**(2), 183–196 (2017). https://doi.org/10.2217/imt-2016-0112
79. Bagheri, E., et al.: Silica based hybrid materials for drug delivery and bioimaging. J. Control. Release **277**(January), 57–76 (2018). https://doi.org/10.1016/j.jconrel.2018.03.014
80. Abbaszad, A., et al.: A Smart pH-responsive Nano-Carrier as a Drug Delivery System: A hybrid system comprised of mesoporous nanosilica MCM-41 (as a nano-container) & a pH-sensitive polymer (as smart reversible gatekeepers): Preparation, characterization and in vitro release st. Eur. J. Pharm. Sci. **93**, 64–73 (2016). https://doi.org/10.1016/j.ejps.2016.08.005
81. Zhang, Y., et al.: Mesoporous silica nanoparticles for increasing the oral bioavailability and permeation of poorly water soluble drugs. Mol. Pharm. **9**(3), 505–513 (2012). https://doi.org/10.1021/mp200287c
82. Sliwinska-Bartkowiak, M., et al.: Melting/freezing behavior of a fluid confined in porous glasses and MCM-41: Dielectric spectroscopy and molecular simulation. J. Chem. Phys. **114**(2), 950–962 (2001). https://doi.org/10.1063/1.1329343
83. Zhao, Q., et al.: Fabrication of mesoporous hydroxycarbonate apatite for oral delivery of poorly water-soluble drug carvedilol. J. Non. Cryst. Solids **358**(2), 229–235 (2012). https://doi.org/10.1016/j.jnoncrysol.2011.09.020

84. Hu, Y., et al.: Incorporation of indomethacin nanoparticles into 3-D ordered macroporous silica for enhanced dissolution and reduced gastric irritancy. Eur. J. Pharm. Biopharm. **79**(3), 544–551 (2011). https://doi.org/10.1016/j.ejpb.2011.07.001
85. Benezra, M., et al.: Multimodal silica nanoparticles are effective cancer-targeted probes in a model of human melanoma. J. Clin. Invest. **121**(7), 2768–2780 (2011). https://doi.org/10.1172/JCI45600
86. Kwon, O.S., et al.: Dual-color emissive upconversion nanocapsules for differential cancer bioimaging in vivo. ACS Nano **10**, 1512–1521 (2016). https://doi.org/10.1021/acsnano.5b07075
87. Zhang, S., et al.: Facile fabrication of dendritic mesoporous SiO 2 @ CdTe @ SiO 2 fluorescent nanoparticles for bioimaging. Part. Part. Syst. Charactarization **33**, 261–270 (2016). https://doi.org/10.1002/ppsc.201500254
88. Walczyk, D., et al.: What the Cell 'Sees' in Bionanoscience. J. Am. Chem. Soc. **132**, 5761–5768 (2010). https://doi.org/10.1021/ja910675v

Ballistic Performance of Nanostructured Armors

Alexander Braun Dresch and Janio Venturini

Abstract Ceramics are widely used as the first layer of protection in many ballistic armors. However, the brittle behavior of these materials can still be the bottleneck in this type of application. One of the possible solutions for this problem is the development of nanostructured ceramic protections. This chapter will discuss the use of nanostructured ceramic armor as a way to improve ballistic protections. In addition, this chapter will address some aspects of the static mechanical properties that are influenced by nanoscale control and what role these properties play in dynamic ballistics. Finally, the equations to evaluate the ballistic performance and the energy dissipation capacity of the ceramic armor will be presented.

Keywords Armors · Advanced ceramics · Ballistic protections · Nanostructure

Abbreviations

MAS	Multilayered armor system
YSZ	Stabilized tetragonal zirconia
ZTA	Zirconia toughened alumina
DOP	Depth of penetration

1 Introduction

Ceramic materials are excellent for ballistic armor applications, mainly due to their mechanical properties, as well as their lower density compared to other materials [1–5]. Therefore, the choice of ceramic material depends on a combination of the various mechanical properties, in addition to several criteria such as resistance to

A. B. Dresch (✉) · J. Venturini
Universidade Federal do Rio Grande do Sul, Porto Alegre, Brazil
e-mail: Alexander.dresch@ufrgs.br

A. Kopp Alves (ed.), *Technological Applications of Nanomaterials*,
Engineering Materials, https://doi.org/10.1007/978-3-030-86901-4_6

damage, energy absorption capacity, weight and resistance to multiple impacts [1, 6–8]. On the other hand, the brittle behavior of the ceramics may be a limiting factor [9], as well as many of its static mechanical properties that are directly related to dynamic ballistic performance [1]. Thus, the main required properties are hardness, fracture toughness, flexural strength and elastic modulus [10, 11].

Just as the destruction capacity of projectiles has been evolving over the years, there is also a need to evolve systems that can guarantee protection. The use of nanostructured ceramics can be a way to improve the capacity of existing materials and overcome limitations inherent to ceramic materials [12]. Therefore, the required mechanical properties of the ceramics can be improved through nanostructural control or nanotechnology. Consequently, a better ballistic performance can be achieved in the application of these advanced ceramics.

2 Nanostructured Ceramic Armors

Ballistic ceramics are widely applied in multilayered armor systems (MAS). These composites use fiber, metal or another resistant material as backing plate and ceramic as the front plate. Thus, the ceramic armors are the first barrier, protecting against high-velocity and high-kinetic energy projectiles [13–15]. As an advantage, ceramic armors present mechanical resistance equivalent to metals, in addition to being much lighter [16]. As a result, compared to other materials for the same purpose, ceramics are excellent for ballistic application mainly due to their low density and excellent mechanical properties [3–5]. However, the brittle behavior of the ceramics translates into great degradation during the impact, providing a reduced capacity to withstand multiple hits [9]. Indeed, this is the major drawback of ceramics in ballistic applications.

One of the ways to circumvent or reduce the limitations of ceramics in these extreme applications is the use of nanostructured ceramic armors. There are two routes for the synthesis of nanostructures in ceramics: physical and chemical methods [12]. However, nanostructured ceramic armors are generally obtained via two different options; either by nanocomposites, through the addition of nanotubes that act as nanofibers in the ceramic matrix, deflecting and improving the fracture toughness [17]; or doing a (physical) structural control of the ceramic matrix on a nanoscale, with grain sizes smaller than 100 nm. The second option presents the best results in mechanical properties compared to the first one. However, the literature on nanostructured ceramics is limited, as they have yet not been able to be produced in an industrial scale. On the other hand, nanocomposites with nano inclusions added in the ceramic matrix have been increasingly used in ballistic armors [18].

Among the many advantages of nanostructured ceramics, the fact that a greater fraction of atoms is found along the grain boundaries is most remarkable. This presence of excess grain boundaries results in exceptional properties that are highly influenced by grain size. Therefore, the mechanical properties such as hardness and fracture toughness are different from their submicron sized counter parts, as dictated by the Hall–Petch effect. When the grain size is very small, the number of dislocations that can be accommodated within the grain is also reduced. Consequently, the energy needed to promote the propagation of dislocations to the neighboring grain is also greater. Lastly, nanostructured ceramics, such as yttria-stabilized tetragonal zirconia (YSZ) used in zirconia-toughened alumina (ZTA) ceramics, provide a way to overcome the brittle nature of these materials. Thus, a nano-ZTA composite is a potential material for ballistic armor applications [18].

3 Aluminum Oxide (Alumina)

Alumina is the most widely used ceramic in the field of ballistic armors [19]. Its high stability and strong atomic bonds give alumina excellent mechanical properties, such as high strength, high modulus of elasticity, high hardness, and good wear resistance, as well as interesting electrical, optical, thermal, chemical, and biological properties [20–29]. However, in contrast to the abovementioned properties, alumina has low fracture toughness and poor flexural strength, as well as low resistance to thermal shock, being more sensitive than silicon nitride and silicon carbide ceramics, for example [30]. In addition, the properties of alumina are highly dependent on processing parameters, sintering temperature, impurity content and grain size, thus resulting in a wide variation in its properties, as shown in Table 1.

3.1 Crystalline Structures of Alumina

Aluminum oxide has several allotropic forms; however, the most stable phase is α-alumina, also called corundum, which is predominantly ionic. The α-alumina (Fig. 1) has a rhombohedral crystal structure in a hexagonal packing network with two thirds of the octahedral sites symmetrically occupied by Al^{3+} cations surrounded by six equidistant O^{2-} ions [31, 32]. The strong interatomic bonding forces between Al^{3+} and O^{2-} ions, partly ionic and partly covalent, give alumina its excellent properties

Table 1 Properties of alumina ceramics

Properties	Sintered
Density (g/cm^3)	3.6–3.95
Sonic velocity (km/s)	9.5–11.6
Flexural strength (MPa)	200–400
Young's modulus (GPa)	300–450
Fracture toughness (MPa m$^{1/2}$)	3.0–4.5
Hardness (GPa)	12–18

Reprinted from Dresch et al. [1] with permission from Elsevier

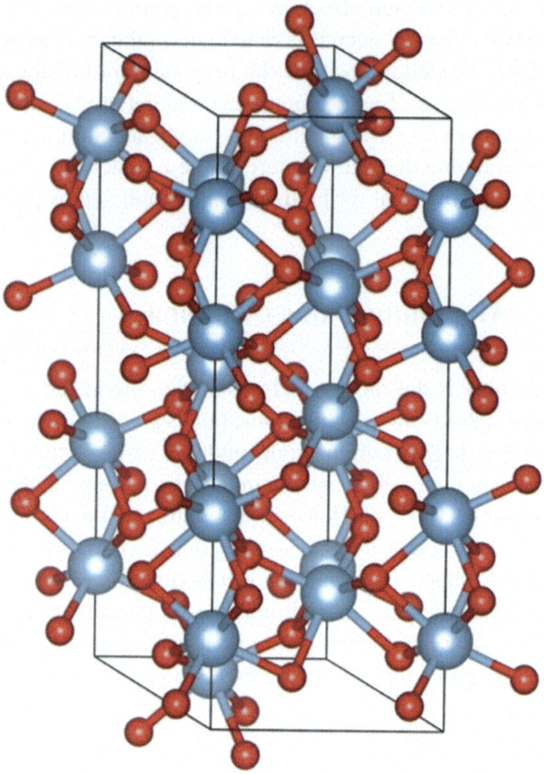

Fig. 1 Crystalline structure of α-alumina

such as physical stability, high melting point (2050 ºC), high hardness, high mechanical resistance, wear resistance, and the advantage of not undergoing oxidation [30, 33, 34].

Moreover, the synthesis of alumina on a nanometric scale usually results in the γ-Al_2O_3 phase. However, even though the γ-Al_2O_3 powder can be made on a nanometric scale, during densification (at approximately 1050 ºC), the γ-Al_2O_3 phase will be transformed into α-Al_2O_3. Thus, this process results in an excessive growth of the grains, losing the characteristics of the nanoscale grain size. Furthermore, the accelerated growth of the grain hinders the densification, which results in a relative density of 82% in conventional sintering without pressure. Therefore, more research involving the processing of nanoscale alumina is required [18].

3.2 Effect on Grain Size on Properties

Hardness is a very important property for ballistic protections, and it can be defined as the superficial resistance that a material presents to the mechanical penetration of another body. Hardness can be influenced by point defects and dislocations, crystallographic structure, and interactions between grain boundaries that restrict free dislocations, reducing plastic deformation. The concentration of stresses that occurs at the grain boundary is lower when the grain size is smaller. therefore, a reduced grain size will result in greater hardness. However, in much smaller grain sizes, this influence will disappear. Thus, when the grain size is reduced to the nanoscale, the relationship between grain size and hardness is that of inverse Hall–Petch; the dislocation glides are no longer the plastic deformation mechanism [18, 35].

Another important property for ballistic protections is fracture toughness, which can be described as the energy absorbed before a material fractures. Grain size plays an important role in this property, as it influences the fracture mode and fracture stress. The decrease in grain size, even at the nanoscale, provides a considerable increase in fracture toughness, indicating that grain refinement is a beneficial option to improve the fracture resistance of ceramic materials. As the grain size decreases, the fracture mode changes. Decreasing the grain size reduces the length of the dislocations, increasing the energy required for dislocations in the grain boundary to occur, thus cleavage fracture toughness increases. However, the grain boundaries act as sources of stress raiser in the material, thus, the intergranular cracks become less tortuous than the transgranular cracks due to the decrease in grain size, decreasing the intergranular fracture toughness. To summarize, larger grain sizes tend exhibit transgranular fracture and smaller grain sizes tend exhibit intergranular fracture. Lastly, a reversed Hall–Petch dependence on the toughness results may occur. As there is a relationship between the crack size and the grain size, even if the crack size is larger than the grain size, the size of the plastic zone at the tip of the crack should correspond to the grain size of the material [18, 36–38].

4 Zirconia-Toughened Alumina (ZTA)

As previously mentioned, the brittle nature of ceramics can compromise the ability to withstand multiple impacts in ballistic applications. Thus, the inherent properties of zirconia-toughened alumina (ZTA) are a way of circumventing the limitations of low fracture toughness.

The introduction of tetragonal zirconia particles in the ceramic matrix of alumina provides an improvement in the mechanical properties, mainly the fracture toughness in comparison with the base alumina. These alumina–zirconia (ZTA) composites are widely used for structural applications, in addition to components for the aerospace industry such as rotors, rocket exhaust cones, cylinders and orifices. The improvement in the mechanical properties of alumina occurs due to the expansion of the zirconia particles in the ceramic matrix upon mechanical loading, causing the martensitic transformation of the tetragonal to monoclinic phase [39, 40]. This phenomenon generates microcracks in the alumina matrix around the zirconia particles. These cracks Locally dissipate the energy, hindering the crack propagation. Thus, for a crack to propagate, it is necessary to apply a higher external tension, which results

Table 2 Properties of zirconia toughened alumina ceramics

Proprieties	Sintered Al_2O_3/3Y-ZrO_2
Density (g/cm^3)	3.4–6.0
Sonic velocity (km/s)	–
Flexural strength (MPa)	280–1000
Young's modulus (GPa)	205
Fracture toughness (MPa m$^{1/2}$)	3.4–9.0
Hardness (GPa)	14–28

Reprinted from Nono et al. [42]

in an increase in fracture toughness and, consequently, in the flexural strength of the alumina–zirconia ceramic composite [40, 41]. Table 2 presents the properties of the ceramic composite, which shows an increase in mechanical properties compared to the values of pristine alumina (Table 1).

4.1 Zirconia Crystalline Structures and Toughening Mechanisms

Zirconia has three allotropic forms, namely, monoclinic (stable below 1,170 °C), tetragonal (stable between 1,170 and 2,370 °C) and cubic °C (stable from 2,370 °C to its melting point at 2,680 °C) [43, 44]. Figure 2 shows a schematic representation of the unit cell of each of these phases.

In the transformation of the tetragonal to monoclinic phase, there is an increase in volume of 3 to 5%, causing a great variation in the crystalline structure. The transformation is of the martensitic type and occurs in a temperature range, which can cause internal stresses and fracture the material. Therefore, the use of pure ZrO2 in many engineering applications is avoided [43–45]. Therefore, the use of additives such as MgO, CaO, Y2O3, and Ce2O3 allow the stabilization of the tetragonal

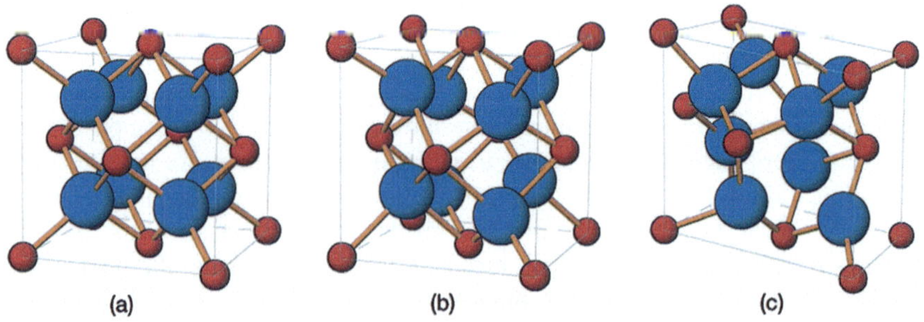

Fig. 2 Representation of the three allotropic forms of ZrO_2: **a** cubic, **b** tetragonal and **c** monoclinic. Reprinted from Mazerolles et al. [45] with permission from John Wiley & Sons

and cubic phases at room temperature. However, the best relationship between the desirable mechanical properties is obtained using multiphase zirconia, known as partially stabilized zirconia (PSZ) [46, 47]. A representation of this stabilization can be seen in Fig. 3, which shows the position of the yttrium ions in the cubic crystalline structure of zirconia [48].

The change from tetragonal to monoclinic phase in partially stabilized zirconia can be induced by surface treatment, temperature and more commonly by stresses, i.e., applying an external load. In this way, this phenomenon provides an improvement in mechanical properties via two different mechanisms [43, 45]. In the first mechanism, the change from the tetragonal to the monoclinic phase occurs due to the propagation of a crack, as seen in Fig. 4. The stresses imposed by the crack propagation reduce the compression of the matrix on the particles of tetragonal zirconia, favoring the transformation to the monoclinic phase. This transformation, associated with a volumetric expansion of 3 to 5%, results in a compressive deformation in the ceramic matrix. Due to this phenomenon being related to the propagation of a crack, extra energy is needed for the crack to continue propagating through the material's nanostructure, thus resulting in an increase in fracture toughness and rupture stress of the ceramic [39, 49].

In the second mechanism, the transformation from tetragonal to monoclinic phase induces micro fissures in the ceramic matrix around the zirconia particles (Fig. 5), causing a reinforcement by microcracks. These microcracks have the ability to absorb part of the energy associated with a crack in propagation, hindering its advancement, resulting in an increase in the toughness of the ceramic [51, 41].

In addition to the mechanisms described above, the zirconia grain size in ZTA ceramics also affects the mechanical properties. The larger the grain size, the lower the tension required to induce the transformation of the tetragonal to monoclinic phase. Similarly, smaller grains are more difficult to transform, requiring a higher load [18].

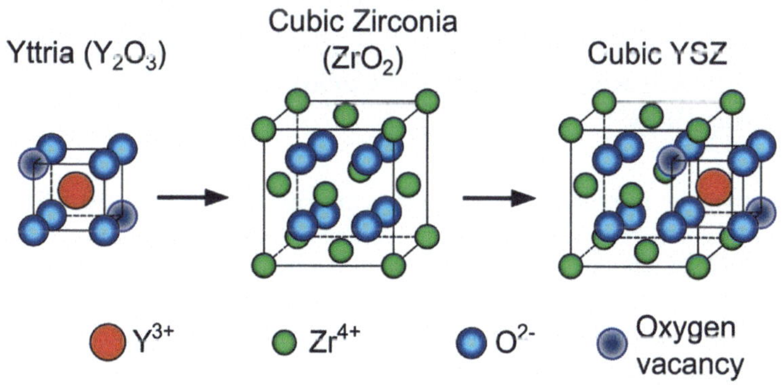

Fig. 3 Representation of the stabilization of cubic zirconia by the addition of yttria

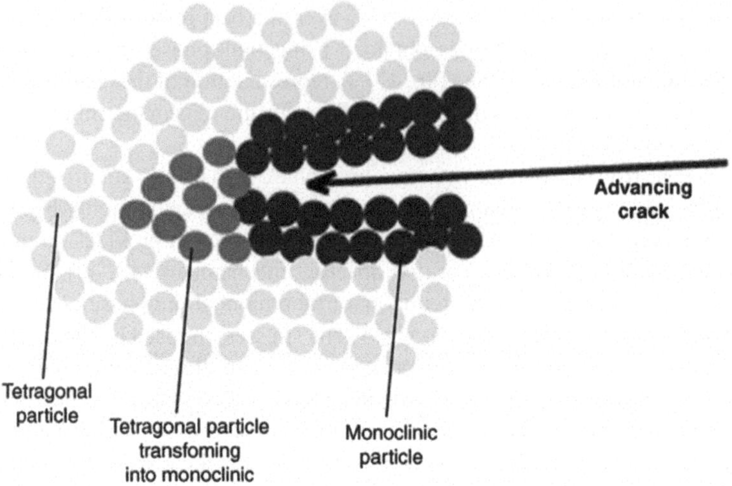

Fig. 4 Representation of the toughening mechanism by transforming the tetragonal to monoclinic phase in zirconia ceramics. Reprinted from Affatato et al. [50] with permission from Elsevier

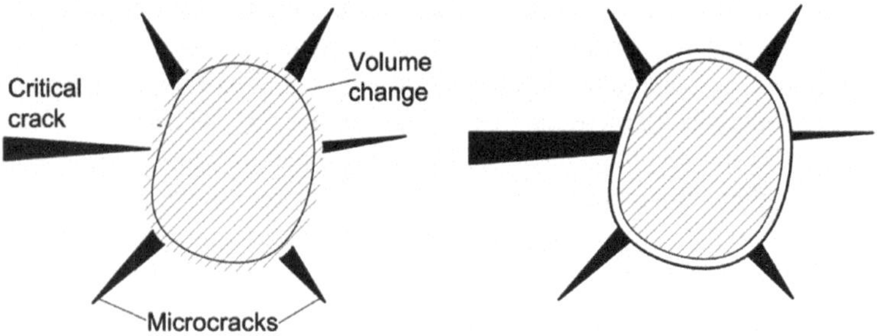

Fig. 5 Reinforcement by the presence of microcracks around a zirconia particle

5 Mechanical Requirements for Ceramic Armor Applications

It is quite complex to establish the necessary requirements for ballistic ceramics, especially when comparing properties obtained in static tests with the dynamic performance of these materials. In addition, due to the dynamic nature of the event, the performance of ballistic ceramics cannot be related to just a single property. Thus, Table 3 presents a list of static properties and their role during a ballistic event. It is worth mentioning that the nano/microstructure of ceramic materials directly affects all properties listed in the table and, consequently, the ballistic performance. Thus, the control of the nano/microstructure is one of the ways for the development of high-performance ballistic ceramics. The correct choice ofmanufacturing processes, such as solid-state sintering, sintering via liquid phase, and hot-press sintering, will

Table 3 Material properties and their role in ballistic performance

Property	Effect in ballistic performance
Nano/Microstructure	Affects all properties listed in the left-hand column below
Grain size	
Minor phases	
Phase transformation or amorphization (stress-induced)	
Porosity	
Density	Weight of the armor system
Hardness	Damage to the projectile
Elastic modulus	Stress wave propagation
Flexure Strength	Multi-hit resistance
Fracture toughness	Multi-hit resistance, field durability
Fracture mode (inter vs. trans granular)	Energy absorption

Reprinted from Karandikar et al. [52] and Dresch et al. with permission from John Wiley & Sons and Elsevier

influence the appearance of secondary phases and porosity. Therefore, these factors have an extreme impact on ballistic performance, as in the case of porosity, which causes a large reduction in the elastic modulus. Another important feature is the grain size, since hardness and flexural strength, as well as the fracture toughness, decrease with the increase in the grain size. Therefore, an appropriate nano/microstructural control is of paramount importance to achieve the required properties [1].

5.1 Hardness, Flexural Strength and Fracture Toughness

During ballistic impact, the first requested property is hardness, which has the function of fracturing and eroding the projectile, preventing penetration and increasing the contact area between the projectile and the ceramic, reducing the impact tension and dissipating the energy. Therefore, the hardness of the ceramic must be greater than that of the projectile [1, 2]. In addition, fracture toughness and flexural strength are very important properties, as they are related to the collapse of ceramic protection, limiting the amount of kinetic energy absorbed during impact. Several studies indicate that the failure mechanism in ceramics subjected to ballistic impacts is due to flexural stresses. Therefore, a high flexural strength means a greater stress load necessary for the rupture to occur. Thus, the greater the flexural strength and the fracture toughness of the ceramics, the greater their capacity to absorb the kinetic energy and resist multiple impacts, which is a necessary requirement to validate the ballistic level according to various standards worldwide [1, 53–57].

5.2 Elastic Modulus and Fracture Mode

In addition to the mentioned properties, a high elastic modulus is also necessary, as the ballistic impact induces high flexural stresses in the ceramic body. Thus, a higher elastic modulus improves the ballistic resistance as a function of the contribution of stiffness in bending deformation. Furthermore, higher elastic moduli result in a longer dwell phase which prolongs the projectile's interaction with the ceramic surface. Finally, the fracture mode is directly related to the amount of energy absorbed by the ceramic body during the impact. In microstructured conventional ballistic ceramics, the transgranular fracture requires more energy to propagate, due to the more tortuous path through the grain boundaries, compared to the intergranular fracture. In nanostructured ceramics, due to the extreme decrease in grain size, more energy is needed for the propagation of the transgranular fracture. The tendency for nanostructured materials is that the intergranular fracture occurs, mainly because the propagation of the crack is less tortuous in this crack mode [1, 18, 36–38].

5.3 Density

Weight is one of the main limiting factors of ballistic ceramics. Heavy personal protection causes loss of mobility, which can have serious consequences for users during operations. The use of heavy protections can also cause joint stresses, in addition to long-term musculoskeletal damage. Therefore, several studies are focused on reducing density and, consequently, the weight of ceramic protections. Thus, it is very important to determine a relationship between the thickness, density and performance of ballistic ceramics [1, 2].

6 Ballistic Performance Evaluation

There are several methods available to assess the performance of ballistic ceramics subjected to the dynamic impact of projectiles. Among all techniques, the depth of penetration (DOP) test is the most adopted. This test may be used as an initial screening, mainly due to its excellent cost–benefit ratio, which entails fast and cheap data collection. However, the collected data does not include the accumulation of damage, stress or tension, even though it is an effective method for assessing the relative performance of the ceramics. Figure 6 shows a schematic diagram of the DOP test. The test consists in measuring the depth of penetration of the projectile into a backing plate with and without ceramic protection. Ballistic performance is analyzed by comparing the difference between the residual moment of the projectile shot at the unprotected backing and the backing protected by ceramics [1].

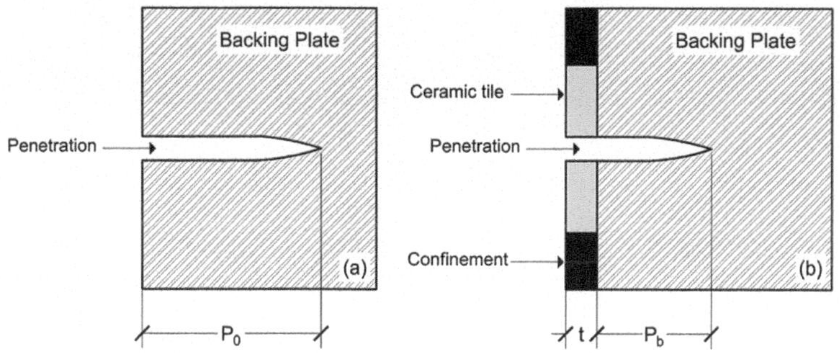

Fig. 6 Schematic diagram of the depth of penetration test configuration: **a** Reference DOP in backing plate without ceramic tiles (P_0); **b** Residual DOP in backing plate (P_b). Reprinted from Dresch et al. [1] with permission from Elsevier

From the data collected in the DOP test, it is possible to evaluate the relative performance of the ceramics using different equations. The mass efficiency (E_m) can be determined through Eq. 1. Depending on the thickness and strength of the ceramic, the reduction of residual DOP increases the mass efficiency. However, increasing the thickness may lead to a decrease in efficiency due to the increase in weight. Equations 2, 3, and 4 define the thickness efficiency (E_{eq}), mass efficiency (M_{eq}), and ballistic efficiency factor (q^2), respectively. The thickness efficiency (E_{eq}) and mass efficiency (M_{eq}) values correlate the performance of the ceramic with the backing plate. The value of E_{eq} defines the thickness of the ceramic material that replaces the thickness of the backing plate to obtain the same protection. M_{eq} represents the ceramic mass that replaces the mass of the backing plate to have the same effect on the projectile. The ballistic efficiency factor q^2 presents the relationship between thickness and mass, given that the dimensions and weight are critical considerations in a protection project. The differential efficiency factor (DEF) can be determined by Eq. 5. This parameter is another method for analyzing mass efficiency. It presents the same results as in Eq. 3, that is, it shows the same linear relationship between the thickness of the ceramic and the residual penetration. Through the linear relationship of the ceramic thickness with the residual penetration, it is possible to define the critical thickness (t_{crit}) (Eq. 6), which represents the maximum thickness of the ceramic so that there is no penetration in the backing plate [1].

$$E_m = \frac{\rho_b \times P_o}{(\rho_c \times t) + (\rho_b \times P_b)} \tag{1}$$

$$E_{eq} = \frac{P_0 \times P_b}{t} \tag{2}$$

$$M_{eq} = \frac{P_0 \times P_b}{t} \times \frac{\rho_b}{\rho_c} \tag{3}$$

$$q^2 = M_{eq} \times E_{eq} \tag{4}$$

$$\text{DEF} = \frac{\rho_b \times (P_0 - P_b)}{\rho_c \times t} \tag{5}$$

$$t_{crit} = \frac{\text{t} \times P_0}{P_o - P_b} \tag{6}$$

where:

ρ_b is the density of the backing plate;
ρ_c is the density of the ceramic tile;
P_o s the reference DOP into the unprotected backing plate,
P_b is the residual DOP into the backing plate after striking the ceramic tile;
t is the ceramic thickness.

6.1 Evaluation of the Ballistic Energy Dissipation

In specific ceramic armors, such as those used in mosaic systems for example, the kinetic energy of the impact must be absorbed through the brittle fracture mechanism of the ceramic. Thus, it is desirable that the ceramic is as brittle as possible, which can be assessed by Eq. 7, where B represents the brittleness of the ceramic. In addition, the figure of merit of the ballistic energy dissipation capacity (D) is represented by Eq. 8. Both expressions have the values of hardness and modulus of elasticity in the numerator and the fracture toughness in the denominator. The lower the fracture toughness and the greater the hardness and elastic modulus, the greater the brittleness of the ceramic and its ability to dissipate the kinetic energy of the impact. However, this type of analysis is not valid for conventional monolithic armor systems, where it is necessary that the ceramic does not have a brittle behavior; therefore, a high fracture toughness is fundamental to withstand multiple impacts. Thus, this criterion should be used to evaluate mosaic armors, where the system is projected to circumvent the brittleness of the ceramics and resist several hits [7, 58, 59].

$$\text{B} = \frac{\text{HV}xE}{K_{IC}^2} \tag{7}$$

$$\text{D} = \frac{0.36(\text{HV}xExc)}{K_{IC}^2} \tag{8}$$

where:

HV is the Vickers hardness;
E is the Young’s modulus;

c is the longitudinal sound velocity;
K_{IC} is the fracture toughness.

7 Conclusion and Future Prospects

Although nano structural control is possibly one of the future options for the development of superior ceramic protections, advances in processing are still necessary to produce these materials on a large scale. Many of the properties required for excellent ballistic ceramics, such as hardness, fracture toughness and flexural strength, are directly influenced by the material's microstructure. Therefore, the control on a nanometric scale would allow the development of improved ballistic ceramics. In addition, each property has a critical role in the performance of ballistic ceramics. Therefore, in order to obtain efficient ballistic ceramics, the combination of several properties is necessary.

The equations presented in Sect. 6 allow us to trace a metrics of ballistic performance in the DOP tests, as well as to use the equations of Sect. 6.1 to meet the ballistic energy dissipation criteria for mosaic armor systems, for example. Finally, nanostructured zirconia-toughened alumina (ZTA) ceramics are a promising possibility, given that they combine the benefits of structuring on a nanoscale with the inherent effects and strengthening of zirconia phase transitions. However, these ceramics have a higher density than pure alumina, and weight is one of the limiting factors in ballistic protections. Therefore, studies are still needed to develop more efficient ZTA ceramics.

References

1. Dresch, A.B., Venturini, J., Arcaro, S., Montedo, O.R.K., Bergmann, C.P.: Ballistic ceramics and analysis of their mechanical properties for armour applications: A review. Ceram. Int. **47**, 8743–8761 (2021). https://doi.org/10.1016/j.ceramint.2020.12.095
2. Dresch, A.B., Venturini, J., Bergmann, C.P.: Improving the flexural-strength-to-density ratio in alumina ceramics with the addition of silicon nitride. Ceram. Int. **47**, 3964–3971 (2021). https://doi.org/10.1016/j.ceramint.2020.09.260
3. Bresciani, L.M., Manes, A., Giglio, M.: An analytical model for ballistic impacts against ceramic tiles. Ceram. Int. **44**, 21249–21261 (2018). https://doi.org/10.1016/j.ceramint.2018.08.172
4. Akella, K.: Studies for improved damage tolerance of ceramics against ballistic impact using layers. Procedia Eng. **173**, 244–250 (2017). https://doi.org/10.1016/j.proeng.2016.12.006
5. Dancer, C.E.J., Spawton, J.N.F., Falco, S., Petrinic, N., Todd, R.I.: Characterisation of damage mechanisms in oxide ceramics indented at dynamic and quasi-static strain rates. J. Eur. Ceram. Soc. **39**, 4936–4945 (2019). https://doi.org/10.1016/j.jeurceramsoc.2019.06.054
6. Yadav, R., Naebe, M., Wang, X., Kandasubramanian, B.: Body armour materials: from steel to contemporary biomimetic systems. RSC Adv. **6**, 115145–115174 (2016). https://doi.org/10.1039/c6ra24016j
7. Medvedovski, E.: Ballistic performance of armour ceramics: Influence of design and structure. Part 1. Ceram. Int. **36**, 2103–2115 (2010). https://doi.org/10.1016/j.ceramint.2010.05.021

8. Medvedovski, E.: Ballistic performance of armour ceramics: Influence of design and structure. Part 2. Ceram. Int. **36**, 2117–2127 (2010). https://doi.org/10.1016/j.ceramint.2010.05.022
9. Akella, K., Naik, N.K.: Composite armour—A review. J. Indian Inst. Sci. **95**, 297–312 (2015)
10. Crouch, I.G., Kesharaju, M., Nagarajah, R.: Characterisation, significance and detection of manufacturing defects in Reaction Sintered Silicon Carbide armour materials. Ceram. Int. **41** (2015). https://doi.org/10.1016/j.ceramint.2015.06.083
11. Hu, D., Zhang, Y., Shen, Z., Cai, Q.: Investigation on the ballistic behavior of mosaic SiC/UHMWPE composite armor systems. Ceram. Int. **43**, 10368–10376 (2017). https://doi.org/10.1016/j.ceramint.2017.05.071
12. Ramaseshan, R., Sundarrajan, S., Jose, R., Ramakrishna, S.: Nanostructured ceramics by electrospinning. J. Appl. Phys. **102**, 111101 (2007). https://doi.org/10.1063/1.2815499
13. Monteiro, S.N., Pereira, A.C., Ferreira, C.L., Júnior, É.P., PondéWeber, R., de Assis, F.S.: Performance of plain woven jute fabric-reinforced polyester matrix composite in multilayered ballistic system. Polymers (Basel). **10**, 1–10 (2018). https://doi.org/10.3390/polym10030230
14. da Luz, F.S., Lima Junior, E.P., Louro, L.H.L., Monteiro, S.N.: Ballistic test of multilayered armor with intermediate epoxy composite reinforced with jute fabric. Mater. Res. **18**, 170–177 (2015). https://doi.org/10.1590/1516-1439.358914
15. Wu, K.K., Chen, Y.L., Yeh, J.N., Chen, W.L., Lin, C.S.: ballistic impact performance of sic ceramic-dyneema fiber composite materials. Adv. Mater. Sci. Eng. **2020** (2020). https://doi.org/10.1155/2020/9457489
16. Haynes, A., Reinhardt, L., Lim, C.: Design and processing of alumina plate composites for ballistic nacre alumina structures. MRS Adv. **3**, 957–962 (2018). https://doi.org/10.1557/adv.2017.631
17. Bolduc, M., Anctil, B., Lo, J., Zhang, R., Lin, S., Simard, B., Bosnick, K., Merati, A.: Ballistic evaluation of nanocomposite ceramic. Pass **2016**, 1–9 (2016)
18. Huang, S., Binner, J., Vaidhyanathan, B.: Nanostructured Advanced Ceramics For Armour Applications. Loughborough University (2013)
19. Rashed, A., Yazdani, M., Babaluo, A.A., Hajizadeh Parvin, P.: Investigation on high-velocity impact performance of multi-layered alumina ceramic armors with polymeric interlayers. J. Compos. Mater. **50**, 3561–3576 (2016). https://doi.org/10.1177/0021998315622982
20. Eftekhari, A., Movahedi, B., Dini, G., Milani, M.: Fabrication and microstructural characterization of the novel optical ceramic consisting of α-Al2O3@amorphous alumina nanocomposite core/shell structure. J. Eur. Ceram. Soc. **38**, 3297–3304 (2018). https://doi.org/10.1016/j.jeurceramsoc.2018.02.038
21. Neuman, E.W., Hilmas, G.E., Fahrenholtz, W.G.: A high strength alumina-silicon carbide-boron carbide triplex ceramic. Ceram. Int. **43**, 7958–7962 (2017). https://doi.org/10.1016/j.ceramint.2017.03.104
22. Lamouri, S., Hamidouche, M., Bouaouadja, N., Belhouchet, H., Garnier, V., Fantozzi, G., Trelkat, J.F.: Control of the γ-alumina to α-alumina phase transformation for an optimized alumina densification, Boletín La Soc. Española Cerámica y Vidr. **56**, 47–54 (2017). https://doi.org/10.1016/j.bsecv.2016.10.001
23. Milak, P., Minatto, F.D., Faller, C., De Noni, A., Klegues Montedo, O.R.: The influence of dopants in the grain size of alumina—A review. Mater. Sci. Forum. **820**, 280–284 (2015). https://doi.org/10.4028/www.scientific.net/MSF.820.280
24. Pawar, M.J., Patnaik, A., Biswas, S.K., Pandel, U., Bhat, I.K., Chatterjee, S., Mukhopadhyay, A.K., Banerjee, R., Babu, B.P.: Comparison of ballistic performances of Al2O3 and AlN ceramics. Int. J. Impact Eng. **98**, 42–51 (2016). https://doi.org/10.1016/j.ijimpeng.2016.08.002
25. McEntire, B.J., Bal, B.S., Rahaman, M.N., Chevalier, J., Pezzotti, G.: Ceramics and ceramic coatings in orthopaedics. J. Eur. Ceram. Soc. **35**, 4327–4369 (2015). https://doi.org/10.1016/j.jeurceramsoc.2015.07.034
26. Ratzker, B., Wagner, A., Sokol, M., Kalabukhov, S., Dariel, M.P., Frage, N.: Optical and mechanical properties of transparent alumina fabricated by high-pressure spark plasma sintering. J. Eur. Ceram. Soc. **39**, 2712–2719 (2019). https://doi.org/10.1016/j.jeurceramsoc.2019.03.025

27. Gajdowski, C., Böhmler, J., Lorgouilloux, Y., Lemonnier, S., d'Astorg, S., Barraud, E., Leriche, A.: Influence of post-HIP temperature on microstructural and optical properties of pure MgAl2O4 spinel: From opaque to transparent ceramics. J. Eur. Ceram. Soc. **37**, 5347–5351 (2017). https://doi.org/10.1016/j.jeurceramsoc.2017.07.031
28. Montedo, O.R.K., Milak, P.C., Faller, C.A., Bergmann, C.P., De Noni, A., Jr.: Effect of LZSA glass-ceramic addition on the erosive wear of pressureless sintered alumina. REM—Int. Eng. J. **73**, 179–188 (2020). https://doi.org/10.1590/0370-44672019730024
29. Montedo, O.R.K., Milak, P.C., Faller, C.A., Peterson, M., De Noni Junior, A.: Effect of LZSA glass-ceramic addition on pressureless sintered alumina. Part II: mechanical behavior. Mater. Res. **21** (2017). https://doi.org/10.1590/1980-5373-mr-2017-0012
30. Smallman, R.E., Bishop, R.J.: Modern physical metallurgy and materials engineering, sixth edit. Butterworth Heinemann (1999). https://doi.org/10.1016/b978-075064564-5/50010-0
31. Auerkari, P.: Mechanical and physical properties of engineering alumina ceramics. VTT Tied.—Valt. Tek. Tutkimusk (1996).
32. Abyzov, A.M.: Aluminum oxide and alumina ceramics (review). Part 1. properties of Al2O3 and commercial production of dispersed Al2O3. Refract. Ind. Ceram. **60**, 24–32 (2019). https://doi.org/10.1007/s11148-019-00304-2
33. Callister, W.D., Rethwisch, D.G.: Materials Science and Engineering—An Introduction, in: Mater. Sci. Eng.—An Introd., 9th edn., Editora LTC, pp. 912 (2016)
34. Wachtman, J.B., Cannon, W.R., Matthewson, M.J.: Mechanical Properties of Ceramics. Wiley, Second Edi (2009)
35. Krell, A., Blank, P.: Grain size dependence of hardness in dense submicrometer alumina. J. Am. Ceram. Soc. **78**, 1118–1120 (1995). https://doi.org/10.1111/j.1151-2916.1995.tb08452.x
36. Armstrong, R.W.: Grain size dependent alumina fracture mechanics stress intensity. Int. J. Refract. Met. Hard Mater. **19**, 251–255 (2001). https://doi.org/10.1016/S0263-4368(01)00046-4
37. Warner, D.H., Molinari, J.F.: Micromechanical finite element modeling of compressive fracture in confined alumina ceramic. Acta Mater. **54**, 5135–5145 (2006). https://doi.org/10.1016/j.actamat.2006.06.046
38. Petch, N.J., Armstrong, R.W.: Work-hardening in cleavage fracture toughness. Acta Metall. **37**, 2279–2285 (1989). https://doi.org/10.1016/0001-6160(89)90155-7
39. Tomaszewski, H.: Effect of sintering atmosphere on thermomechanical properties of Al2O3-ZrO2 ceramics. Ceram. Int. **15**, 141–146 (1989). https://doi.org/10.1016/0272-8842(89)90008-4
40. Evans, K.A.: The manufacture of alumina and its use in ceramics and related applications. Key Eng. Mater. **122–124**, 489–526 (1996). https://doi.org/10.4028/www.scientific.net/KEM.122-124.489
41. Yoshimura, M.: Phase stability of zirconia. Am. Ceram. Soc. Bull. **67**, 1950–1955 (1988)
42. Nono, D.A., Neto, F.P., Mineiro, S.L.: ANÁLISE COMPARATIVA DOS PROCESSOS DE OBTENÇÃO DAS CERÂMICAS DA ALUMINA α, DA ZIRCÔNIA-3YTZP E DO COMPÓSITO ALUMINA α + 18,5 % ZIRCÔNIA-3YTZP PARA APLICAÇÃO EM SATÉLITES COMO BLINDAGEM PARA IMPACTOS COM DETRITOS ESPACIAIS, MINISTÉRIO DA CIÊNCIA E TECNOLOGIA INSTITUTO NACIONAL DE PESQUISAS ESPACIAIS (2016)
43. Saridag, S., Tak, O., Alniacik, G.: Basic properties and types of zirconia: An overview. World J. Stomatol. **2**, 40 (2013). https://doi.org/10.5321/wjs.v2.i3.40
44. Malya, Rao, B., S.R., M., Sajjan, S., AV, R.R.: An overview on zirconia, trends prosthodont. Dent. Implantol. **6** (2015). https://www.researchgate.net/publication/295911625
45. Mazerolles, L., Michel, D., Portier, R.: Interfaces in oriented Al2O3-ZrO2 (Y2O3) eutectics. J. Am. Ceram. Soc. **69**, 252–255 (1986). https://doi.org/10.1111/j.1151-2916.1986.tb07419.x
46. Karihaloo, B.L., Andreasen, J.H. (eds.): Chapter 2 Transformation toughening materials. In: North-Holl. Ser. Appl. Math. Mech. North-Holland, pp. 9–33 (1996). https://doi.org/10.1016/S0167-5931(96)80003-9
47. Zirconia—Handbook of Environmental Degradation of Materials, 3rd edn. (2018)

48. University of Cambridge, YSZ (Zircônia estabilizada com ítria) Estrutura cúbica do tipo fluorita, (n.d.). https://www.doitpoms.ac.uk/tlplib/fuel-cells/sofc_electrolyte.php
49. Ruhle, M., Claussen, N., Heuer, A.H.: Transformation and microcrack toughening as complementary processes in ZrO2-toughened Al2O3. J. Am. Ceram. Soc. **69**, 195–197 (1986). https://doi.org/10.1111/j.1151-2916.1986.tb07405.x
50. Affatato, S., Taddei, P.: Wear phenomena of ceramic joints. In: Wear of Orthopaedic Implants and Artificial Joints, pp. 278–297. Elsevier (2013). https://doi.org/10.1533/9780857096128.1.278
51. Evans, A.G., Faber, K.T.: Toughening of ceramics by circumferential microcracking. J. Am. Ceram. Soc. **64**, 394–398 (1981). https://doi.org/10.1111/j.1151-2916.1981.tb09877.x
52. Karandikar, PG, Evans, G., Wong, S., Aghajanian, M.K., Sennett, M.: A Review of Ceramics for Armor Applications, pp. 163–175 (2009). https://doi.org/10.1002/9780470456286.ch16.
53. Sherman, D., Brandon, D.: Mechanical Properties and their Relation to Microstructure (2008). https://doi.org/10.1002/9783527618217.ch3
54. Association of Test Laboratories for Bullet Resistant Materials and Constructions, VPAM Standard APR 2006 - General basis for ballistic material, construction and product testing, pp. 1–29 (2006)
55. U.S. Department of Justice, NIJ Standard-0101.06—Ballistic Resistance of Body Armor (2008)
56. Russian Federation, GOST 34286–2017—Armored clothing. Classification and general specifications, pp. 1–12 (2017)
57. Associação Brasileira de Normas Tecnicas, ABNT NBR 15000 - Blindagens para impactos balísticos - Classificação e critério de avaliação, pp. 1–10 (2005)
58. LaSalvia, J.C. (ed.): Advances in Ceramic Armor XI. Wiley, New York, Hoboken, NJ, USA (2015). https://doi.org/10.1002/9781119211549
59. Haney, E.J., Subhash, G.: Damage mechanisms perspective on superior ballistic performance of spinel over sapphire. Exp. Mech. **53**, 31–46 (2013). https://doi.org/10.1007/s11340-012-9634-0

Size Effect on Ferroelectricity in Nanoscaled $BaTiO_3$

Lucas Lemos da Silva and Manuel Hinterstein

Abstract Electronic devices are increasingly becoming more present in our everyday life and their relevance will further increase in the future. These devices evolve targeting higher performance, lower power consumption, and better portability. This creates a demand for continuously miniaturized components. Such devices contain many elements based on ferroelectric properties. Barium titanate is the model ferroelectric system. Moreover, its properties are highly dependent on temperature and grain size. It has optimal properties with grain sizes around 1 μm but suffers a sharp dilution when decreasing the grain size. It loses its ferroelectric properties around a critical grain size of 10 nm, or at a nanowire diameter of around 3 nm. The state of the art regarding this subject is presented in this chapter in an introductory and concise manner.

Keywords Barium titanate · Size effect · Ferroelectricity

Abbreviations

BT	Barium titanate
MLCC	Multilayer ceramic capacitor
C	Capacitance
ε_r	Relative permittivity
ε_0	Vacuum permittivity
D	Dielectric displacement
E	External electric field
P_S	Spontaneous polarization
P	Macroscopic polarization

L. L. da Silva (✉) · M. Hinterstein
Karlsruhe Institute of Technology, Karlsruhe, Germany
e-mail: lucas.silva@kit.edu

M. Hinterstein
e-mail: manuel.hinterstein@kit.edu

A. Kopp Alves (ed.), *Technological Applications of Nanomaterials*,
Engineering Materials, https://doi.org/10.1007/978-3-030-86901-4_7

P_{max}	Maximum polarization
P_{rem}	Remanent polarization
E_C	Coercive field
T_C	Curie temperature
OGSR	Optimal grain size range

1 Introduction

Devices based on ferroelectric perovskites are part of our everyday life. Their applications range from medical imaging systems to diesel engine injectors or non-volatile memories, capacitors and thermistors. Barium titanate ($BaTiO_3$ or BT), the protagonist in this chapter, is the model ferroelectric system and its application focuses on multilayer ceramic capacitors (MLCCs). Taking MLCCs as an example, Hong et al. [1] reported in their study that a single smartphone in 2008 contained about 100 MLCCs against 1100 in current smartphones. This surprising increase in the use of these devices is inversely proportional to their size. The constant miniaturization of devices naturally leads to the development and research of material properties in smaller and smaller scales.

In an MLCC the capacitance can be expressed as

$$C = \varepsilon_r \varepsilon_0 \frac{(n-1)A}{d}$$

where A, n and d are respectively the electrode area, number of layers and thickness of each layer. ε_r and ε_0 are the relative and vacuum permittivity, respectively. As the vacuum permittivity is a constant, it is quickly possible to see that capacitors with highest capacity need a large ε_r, small layer thickness and large numbers of layers. However, what is the smallest possible layer thickness? Are we close to the limit?

The scientific community, motivated by the fundamental questions about the existence of a critical size at which the material loses its ferroelectric properties, has been conducting research on the subject for about 50 years [2]–[4]. Currently it is known that in the polycrystalline ceramic BT, this critical size is around 10 nm [5], while for thin films the ferroelectric properties are maintained up to a thickness of 3 nm [6–8].

2 Barium Titanate

Barium titanate is a ceramic material that began to be studied in the mid-1940s. At the time, the high value of its dielectric constant attracted the attention of researchers who soon associated it with ferroelectricity (this topic will be discussed in detail in the next section).

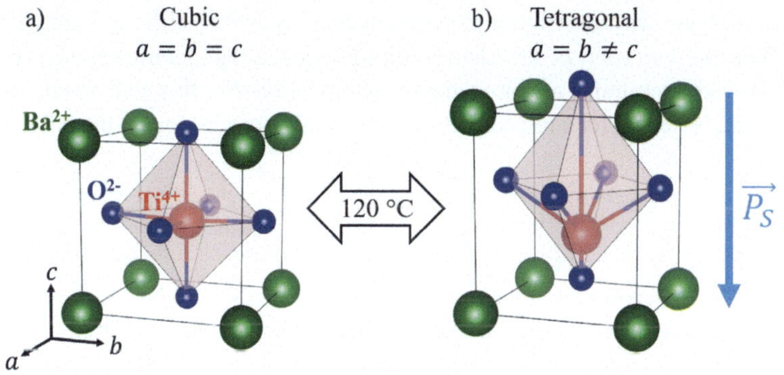

Fig. 1 Perovskite structure of barium titanate. **a** Above the Curie temperature BT crystallizes in the cubic aristotype. **b** Below the Curie temperature a tetragonal distortion of the unit cell occurs. The displacement of the ions lead to the formation of a spontaneous polarisation ($\overrightarrow{\mathbf{P_s}}$)

BT crystallizes in a perovskite crystal structure, in its generic ABO_3 configuration (Fig. 1). The Ba^{2+} cation occupies the A-site in the unit cell vertices and has a coordination number of 12. The Ti^{4+} cation occupies the body-centered B-site in the unit cell and has a coordination number of 6. Above the Curie temperature of 120 °C, its crystal structure possesses cubic symmetry, that is paraelectric and centrosymmetric. When cooled down to room temperature a distortion occurs along the [001] direction, which creates a tetragonal ferroelectric symmetry, where $c/a > 1$. At around 0 °C the structure transforms to orthorhombic symmetry and at around –90 °C to the low temperature rhombohedral phase.

The distortion of the unit cell is accompanied by displacements of the ions. These effects together lead to a spontaneous polarization ($\overrightarrow{P_s}$) along the polar axis of the respective symmetry ($[001]_{pc}$ in tetragonal, $[110]_{pc}$ in orthorhombic and $[111]_{pc}$ in rhombohedral symmetry. The subscript pc denotes pseudocubic indexing). Even though all ions are displaced with respect to each other (Fig. 1b), usually the A-site cation (Ba^{2+}) is set as origin of the unit cell (0,0,0). The B site cation (Ti^{4+}) and the O^{2-} anions are displaced in opposite directions in BT [9]. This increases the spontaneous polarization, which is however small with around 15 μC/cm^2 compared to other typical ferroelectrics such as lead zirconate titanate [10].

3 Ferroelectricity

Ferroelectricity is a characteristic of dielectric materials with crystalline structures that possess a spontaneous polarization which can be reoriented by the application of an external field. The electric dipoles in a ferroelectric crystal occur spontaneously and prior to any external interference. These dipoles are oriented in different directions so that in the material as a whole the resulting polarization can be null.

In a simplistic way, the high dielectric constant in ferroelectric materials is associated with the vibration of the electric dipoles. Since the electric dipoles in these materials occur naturally as spontaneous polarization P_S, this will result in high permittivity. Note that for a ferroelectric, ε_r is directly proportional to the polarization

$$\varepsilon_r \propto \frac{P}{E}$$

where P is the macroscopic polarization and E is the external electric field.

Naturally, an electric dipole interacts with its close neighbors, who tend to align themselves in the same direction. Regions of the crystal, where the electric dipoles are aligned in the same direction, are called ferroelectric domains. The interface separating one ferroelectric domain from another is called a domain wall.

The domains are formed during the transformation from the cubic to the tetragonal phase on cooling. With the distortion of the unit cell, internal stresses arise. To relieve these internal stresses, different regions of the crystal distort in different directions. This leads to complex domain configurations inside the grains. In the tetragonal symmetry only two possible domain walls exist with angles of 90° or 180° between the polarization directions of the domains. Figure 2 shows the ferroelectric domains formed on the polished and etched surface of barium titanate.

It is not possible to write about ferroelectricity without writing about its classic hysteresis loop. Figure 3 shows a typical P-E curve (polarization vs. external electric field). At the point 0, in an unpoled material, the electric dipoles are randomly oriented, resulting in a net zero polarization. Starting from the origin, the path represents the reorientation of the electric dipoles, reacting to the application of the external electric field in one direction, until the resulting polarization reaches a point of saturation, known as the maximum polarization (P_{max}). On removal of the electric field, the electric dipoles do not randomly realign themselves as they did before. Many domains remain aligned parallel to the removed electric field, this is the remanent polarization (P_{rem}). To depolarize the material, it is necessary to apply an electric field

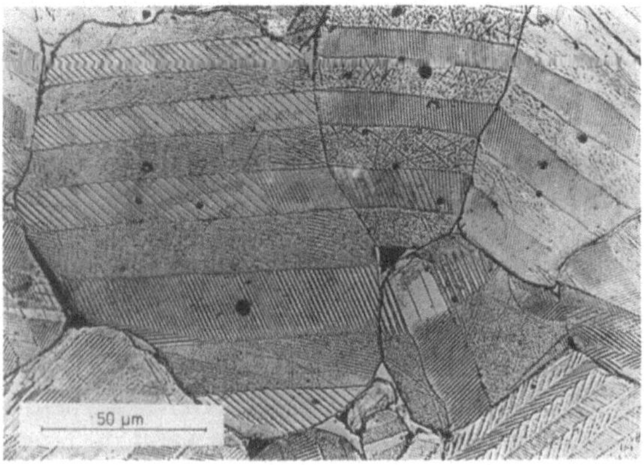

Fig. 2 Vizualization of ferroelectric domains on a polished and etched surface of barium titanate. Reprinted from ref. [11] with the permission from Elsevier

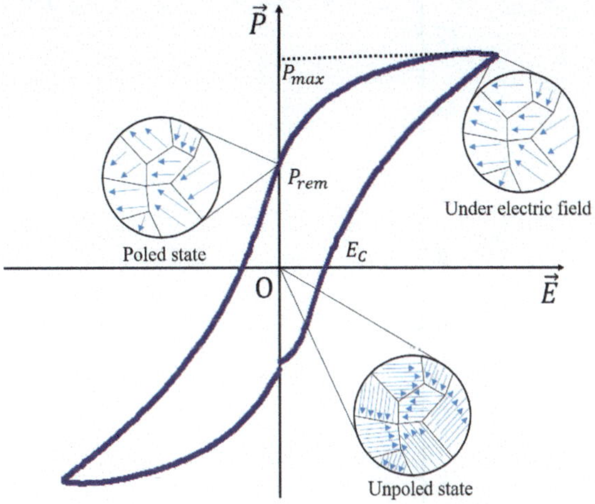

Fig. 3 A typical hysteretic loop of a ferroelectric, which represents the reversibility of its electric dipoles

in the opposite direction with a sufficiently large intensity, known as coercive field (E_C). The electric field can continue to be applied in this direction until saturation. The cycle can be repeated reversibly many times.

4 Size Effect

In the mid-1970s Kinoshita and Yamaji [12] conducted one of the first systematic studies relating the grain size of barium titanate and its dielectric constant. In this work it was presented that at room temperature the dielectric constant of BT varies according to the grain size of the ceramic, reaching the highest values at a grain size of around 1 μm (Fig. 4). The peaks at different temperatures refer to rhombohedral-orthorhombic (–90 °C), orthorhombic-tetragonal (0 °C) and tetragonal-cubic (120 °C) phase transformations [13].

However, at this scale, the variations observed in the ferroelectric properties are not related to the confinement in a small volume, but to microstructural characteristics and several types of interfaces, such as domain walls and grain boundaries [14, 15]. The reduction of the grain size from several microns down to about 1 μm results in increased internal stresses in the material [16], which in turn contribute to the formation of more ferroelectric domains. The mobility of the domain walls as well as the number of domains per volume of material (domain wall density) are crucial for the improvement of the ferroelectric properties. However, while internal stresses help to form ferroelectric domains, they also hinder their mobility. This means that even if the material has a high density of domains, its properties will not be improved if these domains do not have good mobility. In $BaTiO_3$ this optimal point between density of domains and mobility of the domain walls occurs in samples with grain sizes close to 1 μm. Buscaglia and Randall schematically visualized these mechanisms (Fig. 5)

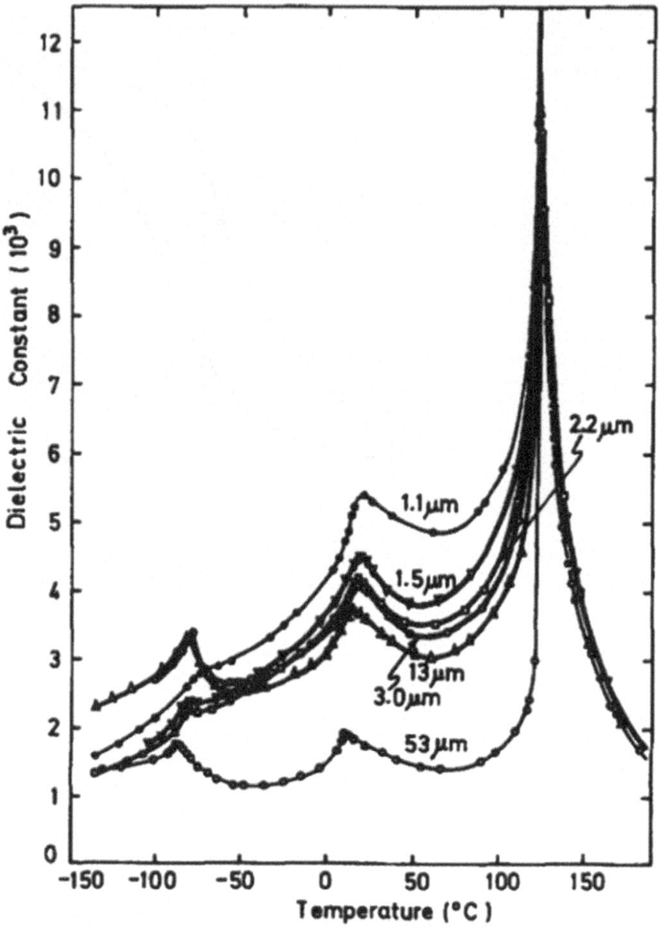

Fig. 4 Dielectric constant in relation to temperature in BT ceramics with different grain sizes. The peaks observed at three different temperatures refer to phase transformations. Reprinted from ref. [12], with the permission of AIP Publishing

[14].

When performing *in situ* studies on BT with intermediate grain sizes, Ghosh et al. [17] detected a phase transformation induced by the electric field, but its symmetry could not be revealed due to a lack of sufficient experimental resolution. More recently, Lemos da Silva et al. [18] investigated the electromechanical mechanisms of BT *in situ* by high-energy X-ray diffraction, combined with a high-resolution detector. Besides clarifying and quantifying the orthorhombic symmetry of the induced phase, their work also revealed the coexistence of orthorhombic and tetragonal phases even in unpoled samples. Near the optimal grain size range (OGSR), where there is high mobility and density of the domain walls, it was verified that the samples that presented the best ferroelectric performance also presented a

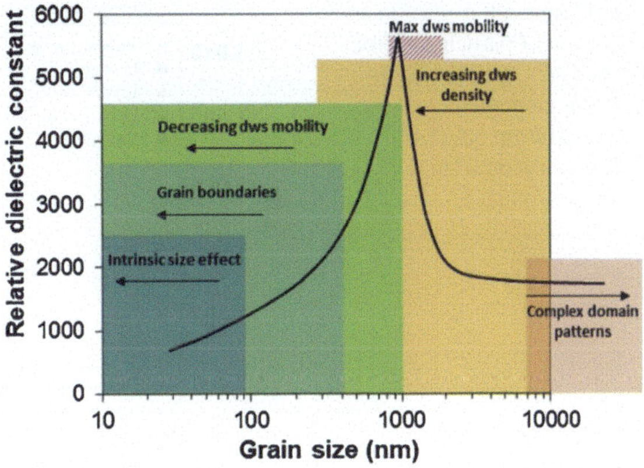

Fig. 5 The mechanisms that can influence the dielectric constant in barium titanate as a function of grain size. Reprinted from ref. [14], with permission from Elsevier

higher fraction of induced orthorhombic phase (Table 1), indicating that this phase transformation plays a fundamental role in the ferroelectric properties of barium titanate.

By reducing the grain size below this optimal point to around 400 nm the mobility of the domain walls is drastically compromised [3, 16, 19]. Frey et al. [4] suggest in their work that the regions near the grain boundary are ferroelectrically inactive. In coarse-grained samples this effect is not very relevant, but in fine-grained BT (100–400 nm) where the volume of interfaces is extremely high, a drastic dilution of the ferroelectric properties occurs. By further reducing the grain size to the nanometer level, the intrinsic size effect dominates and a barium titanate bulk of 10 nm grain size loses its ferroelectric effect at room temperature.

As clarified before, a centrosymmetric structure cannot be ferroelectric e.g. the cubic phase above the Curie temperature is centrosymmetric. Thus, the unit cell distortion (c/a ratio) is a good parameter to evaluate the possibility of the existence of ferroelectricity, which in BT does not occur if $c = a$, typically $c/a = 1{,}011$ in ferroelectric tetragonal BT bulk.

By analyzing the c/a ratio of BT, powders with different particle sizes and synthesized by different techniques (Fig. 6) and different research groups, it can be concluded that this has a strong influence even before the shaping of the material. Böttcher et al. [20] used the metallo-organic precursor pyrolysis technique combined with solid state polymerisation. This technique allows to obtain BT powders with particle size of 10 nm at 600 °C in N_2 atmosphere. Li et al. [21] prepared the powders

Table 1 Summary of electric field induced phase fraction data on BT at two different grain sizes and their respective polarization values [18]

Grain size (μm)	Orthorhombic phase fraction (%)			Polarization (μC/cm^2)	
	Unpoled	2 kV/mm	Remanent	2 kV/mm	Remanent
0.8	46.0 (1.2)	82.3 (1.0)	77.6 (7)	25	12
2.1	42.0 (2)	67.0 (4)	55.1 (3)	18	7

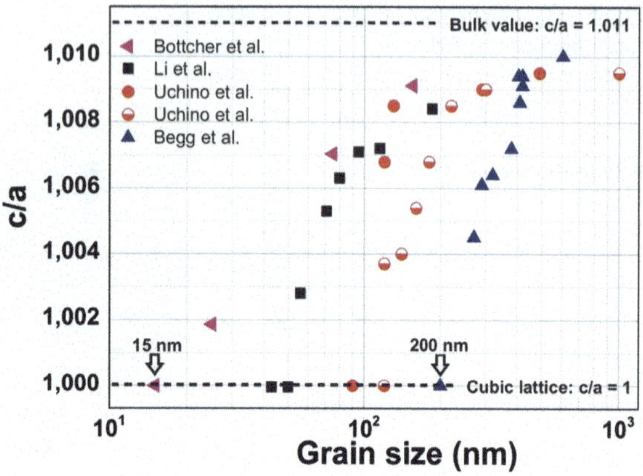

Fig. 6 Unit cell distortion as a function of particle size for BT powders synthesized by different methodologies. Reprinted from ref. [14], with permission from Elsevier

by alkoxide-hydroxide route. This methodology is based on a strong alkaline solution that dispenses calcination, obtaining the powders with particle sizes from 43 nm. Uchino et al. [22] used hydrothermal, coprecipitation and solid state reaction techniques. In a range from 400 to 1100 °C they obtained powders with particle sizes between 90 and 1000 nm. They determined the particle size by X-ray diffraction (full red circle) and by specific surface area (half-full red circle). Begg et al. [23] obtained their powders by a hydrothermal route in a digester furnace for 5 days at 300 °C obtaining powders with 200 nm starting particle size and subsequent firing in a furnace for particle growth. Note that by different synthesis methodologies the critical size (c/a = 1) varied from 15 to 200 nm. This indicates that the defect concentration plays a key role in determining the critical size effect [14].

Besides the influence of the synthesis and processing method, the desired geometry of the material will also influence the critical size. Spanier et al. [24] presented in their study on BT nanowires, that the Curie temperature remains around 120 °C up to 20 nm diameter wires. Below this value the T_C is drastically reduced and almost reaches room temperature in wires with diameters of about 3 nm (Fig. 7). This indicates the loss of ferroelectricity and reveals the critical size for BT with this geometry.

These results show that the long-range ordering of dipoles in the crystal structure determines the structural and dielectric properties in barium titanate. With grain sizes at the OGSR the high density of domain walls, the good mobility (Figs. 4 and 5) and the electric field-induced orthorhombic-tetragonal phase transformation (Table 1) enhance these properties. The reduction of the grain size below the OGSR leads to a drastic dilution of the properties while the number of interfaces increases. At this point, the defects originating from the powder synthesis start to exert a more relevant influence (Fig. 6). Descending to the nanometer scale the intrinsic confinement effect prevails and the shape in which the material is designed will determine the critical size of about 10 nm for bulk, 3 nm diameter for nanowires (Fig. 7) or thickness of thin films.

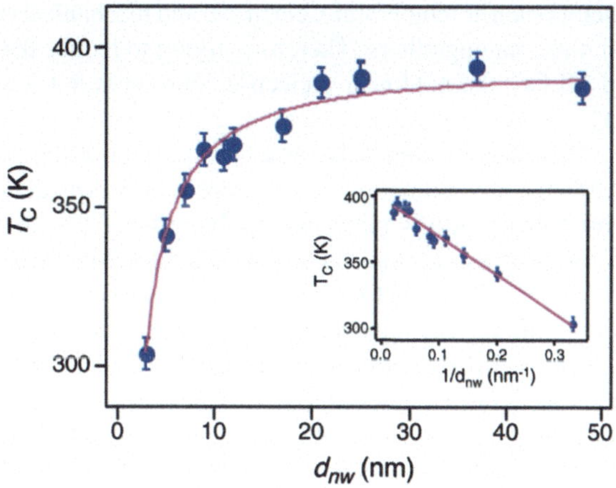

Fig. 7 Ferroelectric-paraelectric phase transformation (T_C) as a function of nanowire diameter (d_{nw}). Reprinted with permission from ref. [24]. Copyright 2021 American Chemical Society

5 Challenges and Perspectives

The most modern MLCCs being manufactured today have a layer thickness of around 500 nm with a grain size around 100 nm and sometimes smaller. Figure 8 is a cross section of an X5R MLCC, with dielectric layers of approximately 450 nm and co-fired nickel electrodes. The dielectric layer grain size is in the range of 80–200 nm.

However, with this grain size, much of the permittivity of the material has already been diluted and it is possible to make more layers and thus compensate for this loss and achieve an optimum capacitance per volume of device. Nevertheless, in addition to the size effects, these devices are often subject to conditions that can be limiting

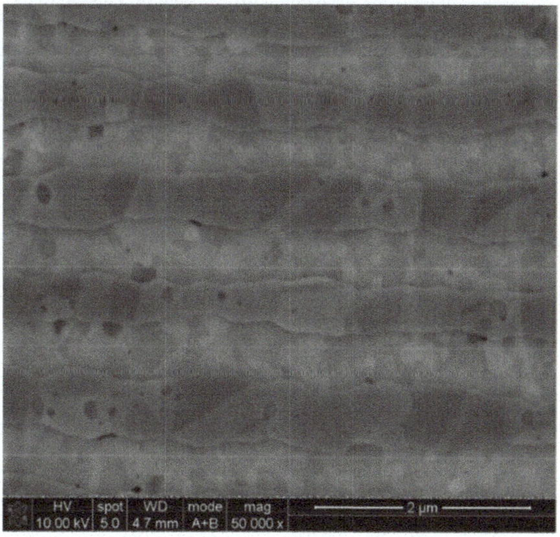

Fig. 8 Cross section of a X5R multi-layer ceramic capacitor. Reprinted from ref. [14], with permission from Elsevier

factors such as temperature, electrical and mechanical stress, electrode interfaces and lifetime, among others. Therefore, trying to reduce the grain size or layer thickness of MLCC, the nonlinear dielectric constant is not the only challenge to overcome [1].

This chapter concisely presents the state-of-the-art on the critical size of BT. Although the optimal properties appear in intermediate grain sizes (0,6–2 μm), the cost benefit ratio is better for applying the BT with small grain sizes, sacrificing ε_r but processing more layers and a smaller final device. As explained above, the commercial X5R MLCCs operate with grain sizes around 80–200 nm and the critical size of the BT is around 10 nm. In the next few years we will be at the physical size limit of the MLCCs, and other ferroelectric devices. Although there is still no practical solution to further reduce the size of these devices, there is still much to be developed regarding the processing of more environmentally friendly ferroelectric devices.

References

1. Hong, K., Lee, T.H., Suh, J.M., Yoon, S.H., Jang, H.W.: Perspectives and challenges in multilayer ceramic capacitors for next generation electronics. J. Mater. Chem. C **7**(32), 9782–9802 (2019)
2. Jaffe: Piezoelectric Ceramics, vol. 3 (1971)
3. Arlt, G., Hennings, D., De With, G.: Dielectric properties of fine-grained barium titanate ceramics. J. Appl. Phys. **58**(4), 1619–1625 (1985)
4. Frey, M.H., Xu, Z., Han, P., Payne, D.A.: Role of interfaces on an apparent grain size effect on the dielectric properties for ferroelectric barium titanate ceramics. Ferroelectrics **206–207**(1–4; 1–2), 337–353 (1998)
5. McCauley, D., Newnham, R.E., Randall, C.A.: Intrinsic size effects in a barium titanate glass-ceramic. J. Am. Ceram. Soc. **81**(4), 979–987 (1998)
6. Lichtensteiger, C., Triscone, J.M., Junquera, J., Ghosez, P.: Ferroelectricity and tetragonality in ultrathin PbTiO3 films. Phys. Rev. Lett. **94**(4), 1–4 (2005)
7. Fong, D.D. et al.: Ferroelectricity in ultrathin perovskite films. Science (80-), **304**(5677), 1650–1653 (2004)
8. Junquera, J., Ghosez, P.: Critical thickness for ferroelectricity in perovskite ultrathin films. Nature **422**(6931), 506–509 (2003)
9. Nakatani, T. et al.: Variable-temperature single-crystal X-ray diffraction study of tetragonal and cubic perovskite-type barium titanate phases. Acta Crystallogr. Sect. B Struct. Sci. Cryst. Eng. Mater. 72(1), 151–159 (2016)
10. Frantti, J., et al.: Phase transitions of Pb(ZrxTi1-x)O3 ceramics. Phys. Rev. B **66**(6), 1–15 (2002)
11. Hennings, D.: Barium titanate based ceramic materials for dielectric use. Int. J. High Technol. Ceram. **3**(2), 91–111 (1987)
12. Kinoshita, K., Yamaji, A.: Grain-size effects on dielectric properties in barium titanate ceramics. J. Appl. Phys. **47**(1), 371–373 (1976)
13. Moulson, A.J.: Electroceramics, vol. 91 (2003)
14. Buscaglia, V., Randall, C.A.: Size and scaling effects in barium titanate. An overview. J. Eur. Ceram. Soc. (November 2019), 0–1 (2020)
15. Ihlefeld, J.F., Harris, D.T., Keech, R., Jones, J.L., Maria, J.P., Trolier-McKinstry, S.: Scaling effects in perovskite ferroelectrics: fundamental limits and process-structure-property relations. J. Am. Ceram. Soc. **99**(8), 2537–2557 (2016)

16. Picht, G., Khansur, N.H., Webber, K.G., Kungl, H., Hoffmann, M.J., Hinterstein, M.: Grain size effects in donor doped lead zirconate titanate ceramics. J. Appl. Phys. **128**(21), 214105 (2020)
17. Ghosh, D., et al.: Domain wall displacement is the origin of superior permittivity and piezoelectricity in BaTiO3 at intermediate grain sizes. Adv. Funct. Mater. **24**(7), 885–896 (2014)
18. da Silva, L.L., Picht, G., Hoffmann, M.J., Hinterstein, M.: Uncovering the symmetry of induced ferroelectric phase transfor-mation in polycrystalline barium titanate. J. Appl. Phys. p. in preparation (2021)
19. Arlt, G., Dederichs, H., Herbiet, R.: 90°-Domain wall relaxation in tetragonally distorted ferroelectric ceramics. Ferroelectrics **74**(1), 37–53 (1987)
20. Böttcher, R. et al.: Size effect in-doped nanopowders observed by electron paramagnetic resonance. Phys. Rev. B - Condens. Matter Mater. Phys. **62**(3), 2085–2095 (2000)
21. Li, X., Shih, W.H.: Size effects in barium titanate particles and clusters. J. Am. Ceram. Soc. **80**(11), 2844–2852 (1997)
22. Uchino, K., Sadanaga, E., Hirose, T.: Dependence of the crystal structure on particle size in barium titanate. J. Am. Ceram. Soc. **72**(8), 1555–1558 (1989)
23. Begg, B.D., Vance, E.R., Nowotny, J.: Effect of particle size on the room-temperature crystal structure of barium titanate. J. Am. Ceram. Soc. **77**(12), 3186–3192 (1994)
24. Spanier, J.E., et al.: Ferroelectric phase transition in individual single-crystalline BaTiO 3 nanowires. Nano Lett. **6**(4), 735–739 (2006)

Electrochromic Nanomaterials

Filipe Ailan da Silveira, Adaiane Parisotto, Felipe Amorim Berutti, and Annelise Kopp Alves

Abstract The wide range of potential applications of electrochromic devices has arisen great interest in the research and development of electrochromic materials. For example, civil construction, automobile, and aviation industries have already put electrochromic windows (smart windows) on the market. These windows can modulate light transmission and thus increase the energy efficiency of buildings and increase comfort to drivers and passengers. Advances in nanotechnology have also allowed electrochromic materials to be more efficient, resulting in new technology applications. This chapter will briefly discuss electrochromism, organic and inorganic electrochromic materials, some electrochromic devices, advantages for their use, and future perspectives for this technology.

Keywords Electrochromism · Thin films · Electrochromic devices · Energy efficiency

1 Introduction

Both researchers and industry have shown a lot of interest in technologies involving electrochromic devices. Not only have these devices become efficient alternatives for reducing energy consumption, but they can also provide environments with greater quality. The correct application of an electric field leads to the color variation between transparent and colored states. In buildings, electrochromic windows control the passage of sunlight via transmittance, determining the transfer of thermal radiation in its interior and allowing an increase of up to 25% in energy efficiency [1, 2]. These "electrochromic windows" are mainly composed of thin films with a substrate of tungsten oxide and nickel oxides, distinguished by a polymeric electrolyte layer or solid inorganic thin film [3, 4].

The first studies appeared in the 60 s and focused mainly on thin films of transition metal oxide, whose chief characteristic was low energy consumption. In 1980, liquid crystal displays (LCDs) gained significant space due to their ability to switch from

F. A. da Silveira (✉) · A. Parisotto · F. A. Berutti · A. K. Alves
PPGE3M, Universidade Federal do Rio Grande do Sul, Porto Alegre, Brazil

A. Kopp Alves (ed.), *Technological Applications of Nanomaterials*, Engineering Materials, https://doi.org/10.1007/978-3-030-86901-4_8

translucent to opaque state very quickly [5, 6]. Electrochromic displays, however, presents an excellent low-power and versatile technology for passive displays.

Organic and inorganic materials, liquid crystalline polymers, and conjugates are key for producing and constructing electrochromic devices [7]. Other factors also influence the efficiency result of an electrochromic device, such as the correct choice of the deposition method, which defines the microstructure, the thickness, and the porosity of the final resulting material [8]. Therefore, this chapter seeks to analyze electrochromism, electrochromic devices, and their applications in further detail.

2 Electrochromism

Electrochromism is a reversible color change in a given material caused by an electric field or a specific current applied to it. It is the reversible and perceptible change in transmittance related to an electrochemical method of an induced reaction of oxidation–reduction. Color variation can typically occur between a transparent and a colored state or even between two colored states. The visual change is achieved by the application of an electric current [9].

Electrochromic devices are rechargeable batteries, where the electrochromic electrode is singled out by a liquid or solid electrolyte that is suitable for a charge balance against the electrode, and color changes occur when the electrochemical cell is charged and depleted with the application of a few volts [9].

With the incidence of electrical potential, electrochromic materials can modify their optical properties [7]. Therefore, when the polarity of the voltage applied is reversed, the changes are reversible, and the material can revert to its original state. Electrochromic materials suffer reversible changes in optical density after oxidation and induced electric voltage drops. Instead of a constant stimulus to maintain emission, devices that include electrochromic materials need only a sufficient initial electrical stimulation to cause a change in the oxidation state and, consequently, in color. These devices, however, are more appropriate for passive applications, such as in information billboards and especially in smart windows, in which the environment is naturally bright, and the chosen color remains for a long time [7, 10, 11].

In electrochromic windows, the passage of sunlight can be modulated by controlling transmittance, which effectively determines the transfer of thermal radiance and improves energy efficiency in buildings [2]. The automotive industry has also been investing in vehicles with anti-reflective mirrors, and some airplanes allow their users to control the entry of light through electrochromic windows [1, 5].

"Smart windows" typically contain thin films with a substrate of tungsten oxide and nickel oxide separated by a solid inorganic thin-film ion transmitter or a layer of polymeric electrolyte. Absorbance modulation occurs when tension is applied between the tungsten film and thin oxide and nickel-oxide films using transparent electrical conductors so that balanced charge electrons and ions are conducted through the two films [3, 4, 12].

Electrochromic materials can be divided into three classes: organic, transition metal oxides, and intercalation materials. Transition metal oxides are the most researched materials. They behave as mixed conductors of electrons and ions and thus require the presence of an electrolyte with the capacity to store and transport these ions [1].

The technology of electrochromic displays has been widely studied. It originated in the mid-1960s and focused initially on thin films of transition metal oxides [5]. One of its characteristics is low energy consumption, requiring a mere initial electrical impulse to change the oxidative state and allow color changes. In the 1980s, liquid crystal displays gained market share due to their ability to quickly switch between translucent and opaque states [6].

Electrochromic materials based on the electrothermal chromatic process include inorganic compounds as well as reduced organic molecules, conjugated polymers, and liquid crystalline polymers [13].

3 Inorganic Materials

When deposited as thin films, certain inorganic compounds can be used in the construction of electrochromic devices. When applying an appropriate electrolyte, the mixed conduction of electrons and ions is possible [7]. During the intercalation/deintercalation of these ions, there is an accompanying flow of electrons in the opposite direction, which modulates the materials' optical properties. If the electrochromic effect causes color formation after the intercalation of ions, it is said to be cathodic. Otherwise, the electrochromism is anodic if the colored state occurs after the deintercalation [1, 14].

3.1 Inorganic Oxide Materials

Among the inorganic electrochromic materials, transition metal oxides are the most widely researched, particularly the tungsten structure of tungsten trioxide (WO_3), which can be represented as corner-sharing WO_6 octahedra, known as an 'empty-perovskite' structure [8] (Fig. 1). The occurrence of some edge-sharing arrangements reflects the tungsten oxide tendency of forming sub-stoichiometric oxides $WO_{3\text{-}X}$. Each of these oxides presents a variation of colors, from yellowish (WO_3) to blue ($WO_{2.9}$), violet ($WO_{2.72}$), and brown (WO_2) [15].

As an electrochromic device, a voltage application induces ions to intercalate interstitial sites of the WO_3 atomic lattice and change its electronic density. This process causes the reversible formation of a tungsten bronze, following the equation: $xM^+ + xe^- + WO_3 \rightleftarrows M_xWO_3 (0 < x < 1)$ [16]. Consequently, some of the tungsten atoms have a valence change from W^{6+} to W^{5+}, resulting in color formation. There is a dependency between the optical effect intensity and the electronic density,

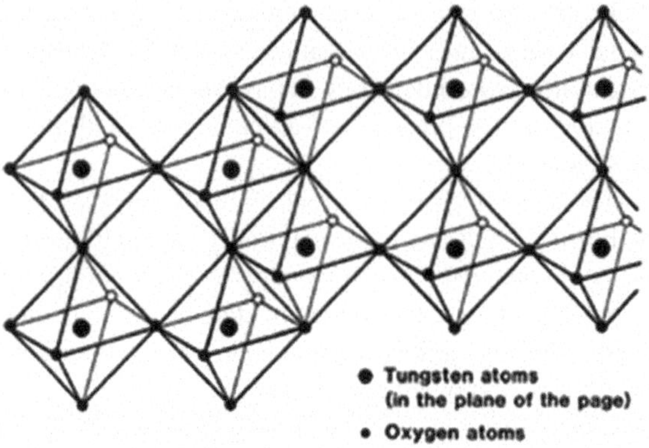

Fig. 1 Representation of corner-sharing and edge-sharing octahedra in slightly substoichiometric crystalline W oxide. Reprinted with permission from Granqvist [8]

and it is proportional to the number of intercalated ions. The effect is also persistent until removing those electrons and intercalated ions, where the tungsten ions return to their initial W^{6+} state [17].

Almost every electrochromic transition metal oxide crystal structure can be described as MeO_6 octahedra frameworks [7]. Considering the perovskite structure of general composition $CMeO_3$, the metal ions (Me) occupy cubic unit cell vertices, while oxygen ions are positioned on the edges (Fig. 2). For electrochromic oxides such as WO_3, $\beta - MoO_3$, and ReO_3, the central atom "C" is not present, for which the 'empty-perovskite' structure is named.

As stated earlier, this structure can be represented as a corner-sharing octahedra framework, where the metallic ion is connected to six oxygen ions. Furthermore, the interstitial sites between each of these octahedra make it possible for mobile

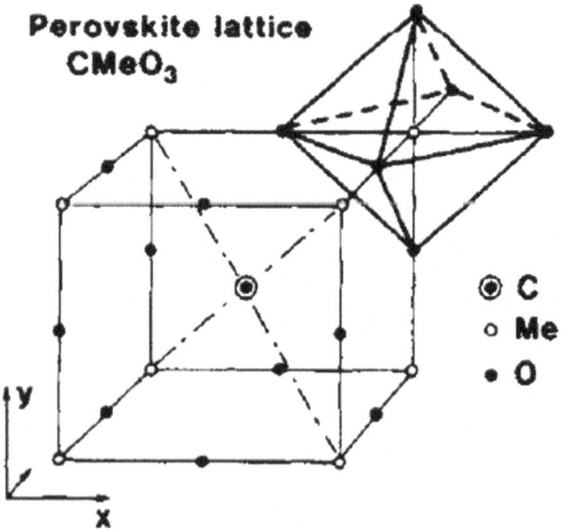

Fig. 2 The unit cell for the perovskite lattice with emphasis on octahedral symmetries. Reprinted with permission from Granqvist [7]

ions to intercalate [17]. Other electrochromic materials of interest are TiO_2, MnO_2, VO_2, RuO_2, IrO_2, and RhO_2. These oxides, differently from WO_3, present a rutile-like structure that can be represented as edge-sharing MeO_6 octahedral units [7]. Similarly, the interstitial sites of its structure allow intercalation of ions.

3.2 *Inorganic Non-Oxide Materials*

Ferric hexacyanoferrate, also known as Prussian Blue, is arguably the most well-known non-oxide inorganic electrochromic material [7]. It belongs to the hexacyanometalates class of materials, which includes compounds of the general formula $M_k[M'(CN)_6]_l$, where M and M' are transition metals of different valencies [18]. Prussian Blue is a ferrocyanate of chemical formula $Fe^{3+}{}_4[Fe^{2+}(CN)_6]_3$ and is frequently used as an inorganic pigment for its characteristic blue color.

Prussian Blue can be applied as electrochromic material due to the loss of its coloration when reduced to $Fe^{2+}{}_4[Fe^{2+}(CN)_6]_3{}^{4-}$, also known as Prussian White [7, 18, 19]. In an anodic electrochromic device, with the intercalation of K^+ ions, for example, a redox reaction occurs, described by the following equation: $Fe^{3+}{}_4[Fe^{2+}(CN)_6]_3(\text{blue}) + 4e^- + 4\,K^+ \rightleftarrows K_4Fe^{2+}{}_4[Fe^{2+}(CN)_6]_3$ (colorless) [20]. Through the substitution of the iron ions, the Prussian Blue has several analogous compounds with similar electrochromic properties and of several different colors [7].

4 Organic Materials

Compared to inorganic electrochromic materials such as WO_3 and other transition metal oxides, organic electrochromic materials may offer some advantages, such as flexible devices, a wider color variety, and renewable source materials [21, 22]. Also, the processing of organic materials usually involves hazardous solvents, though some electrochromic water-soluble organic materials have been recently reviewed [21]. Certainly, the research and development of these water-soluble materials show great potential, considering the increased necessity of environmentally and economically sustainable technologies.

The most common classes of organic electrochromic materials are viologens, metallophthalocyanines, and metallopolymers. [18, 23–25]. There are, besides these classes of materials, many known organic electrochromic materials compounds that will not be covered here, which include carbazoles, methoxy-biphenyls, quinones, pyrazolines, and conducting polymers [25].

4.1 Viologens

Viologens, chemically called 1,1-Di(hydrocarbyl)-4,4'-bipyridinium salts, are organic compounds with the general chemical formula $(C_5H_4NR)_2{}^{n+}$. They were first discovered in 1993 by Michaelis and Hill [26] as oxidation–reduction indicators for biological purposes. Eventually, some of their derivatives, known as paraquats, were used as herbicides [27]. They have also been used as electroactive layers to modify electrode surfaces [28, 29] and as electrode materials for batteries [30].

The 1,1'-substituents influence the solubility and electrochemical properties of viologens. Different substituents have been thoroughly studied concerning the electrochromic behavior of viologen [31]. For example, one of the aims of using organic electrochromic materials is to produce flexible and electrolyte-free devices. The work of Seo et al. [32] reports the development of such device, based on N,N′-dimethyl-4,4′-bipyridinium (MV^{2+}) (methyl viologen) as a cation and conductive graphene quantum dots (GQDs) as an anion. In their work, the electrochemical electron transfer steps of MV^{2+} were analyzed by cyclic voltammetry in a three-electrode electrochemical cell using a GQD solution without an electrolyte, as well as in a well-known electrolyte solution of KCl. As a result, the authors proved to be possible to obtain a stable, flexible, and electrolyte-free ECD using a viologen and GQD.

4.2 Metallophthalocyanines

Metallophthalocyanines (MPcs) are aromatic systems with metal ions (M) at the center of the macrocycle (Fig. 3) [33]. They are structural analogs of porphyrins, and their unique physicochemical properties depend on their molecular packing [34]. MPcs containing transition metals are planar and exhibit intense absorption

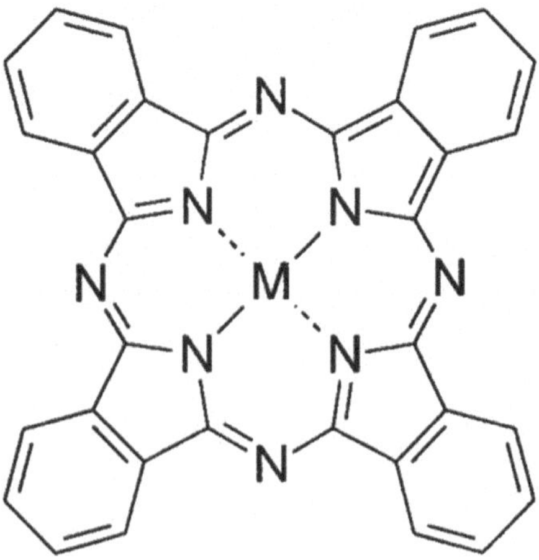

Fig. 3 The structural, general chemical formula of a metallophthalocyanine. Reprinted with permission from Claessens et al. [33]

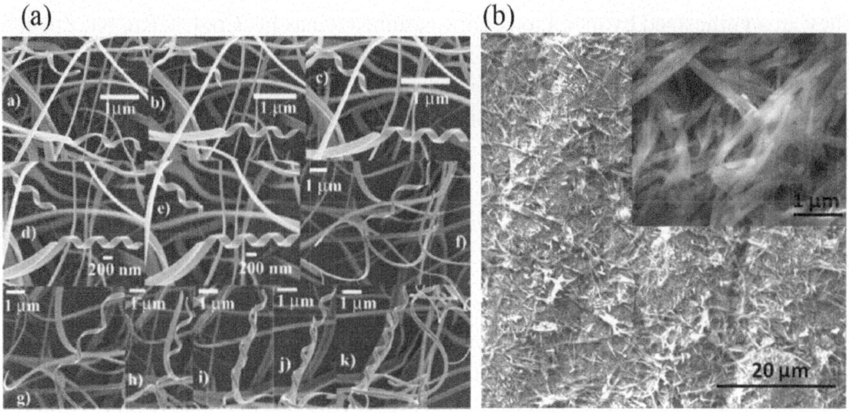

Fig. 4 **a** SEM images illustrating the formation of F16CuPc helices (reprinted with permission of 37) **b** FESEM images of NiPc-NF-coated graphite rod electrode (inset show magnified image). Reprinted with permission from Tong et al. [38]

ranging from ultraviolet to near-infrared region [35]. Although original MPcs are not electrically conductive, the applications often include conductive polymers to enhance their properties.

MPcs can self-assemble and form supramolecular structures through π-π stacking, making them easy to synthesize in an elongated shape. Researchers have fabricated various morphologies of MPcs: Cu and Zn PCs in nanosized micelles to yield spherical nanostructures [36], 3D-flower-like structures [37], different metal phthalocyanines (copper, nickel, iron, cobalt, and zinc), and copper hexadecafluorophthalocyanine (F16CuPc) nanoribbons and nanorods (Fig. 4a) [38] and NiPc coated graphite needle like nanostructures (Fig. 4b) [39].

4.3 *Metallopolymers*

Combining the electrochromic properties of transition metal ions with those of organic molecules and polymers gives rise to another class of electrochromic materials: transition metal complexes or metallopolymers. The electrochromic properties of transition metal complexes are based on the redox reaction of the metal ion or ligand molecule and can be modified by changing their structure [40].

Also known as metallo-supramolecular polyelectrolytes (MEPE), this class of material is prepared by metal ion-induced self-assembly of ditopic bis-terpyridines. The modular nature of the metal ion permits controlling the structure and the properties of MEPE by choosing the metal ions and ligands [41].

Metallopolymers (MEPE) have been investigated as cathodically-coloring (upon reduction and Li^+ insertion) [42, 43]. The metal-to-ligand charge-transfer (MLCT) transition in the visible spectrum is responsible for the color of the metallopolymers.

They are synthesized by metal ion coordination, such as Fe, Co, Cu, Ru, Ni, Zn, and a ditopic ligand such as 1,4-bis(2,2′:6′,2″-terpyridine-4′-yl)benzene [44]. MEPEs show high contrast ratios (TMLCT values from < 5% (dark) to > 80% (bright)), short response times, high cycle stability, and high coloration efficiencies ($\eta > 500$ $cm^2\ C - 1$) [45].

5 Synthesis Methods

Choosing an adequate deposition method during the construction of an electrochromic device is very important since it helps define the microstructure, thickness, and porosity of the resulting thin film. A WO_3 film may be obtained in amorphous or crystalline forms, with a tendency of electrochromic performance improving with the reduction of crystal sizes [8]. Recently, researchers have been focusing on the study of nanostructured thin films, which have demonstrated higher electronic densities and coloring efficiency, given the increase in surface area and ion insertion kinetics.

Some of the commonly used processes for electrochromic device fabrication are listed in Table 1. These can be classified according to the state of used materials and precursors [46].

Table.1 Processes for electrochemical device fabrication

Solution-based	Electrodeposition	
	Sol–gel	Dip coating Spin coating Spraying coating
	Spray pyrolysis[a]	
Vapor based	Thermal evaporation deposition	
	CVD	HWCVD AACVD AP-PECVD oCVD
	Sputtering	DC sputtering RF sputtering Magnetron sputtering Pulsed DC magnetron sputtering Reactive sputtering
Solid particle-based	Nanoparticle deposition system	

[a]Spray pyrolysis is in between solution and vapor based

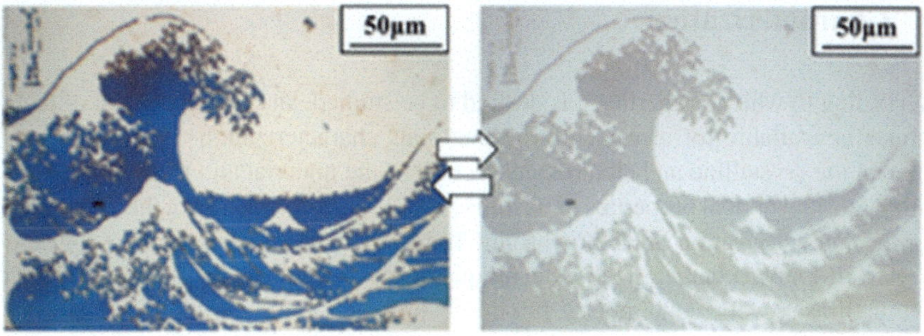

Fig. 5 Optical micrograph of Prussian Blue nanoparticles forming an image representation of "The Great Wave off Kanagawa" by Hokusai, in both colored and uncolored states. Reprinted with permission from Shimojo [20]

Solution-based methods include electrodeposition and sol–gel techniques, such as dip coating, spin coating, and spraying coating. Chemical vapor deposition, sputtering and thermal evaporation depositions are all vapor-based and can be subdivided into different special techniques. Spray pyrolysis may be classified as an intermediate between solution-based and vapor-based since it starts as a solution but, during deposition, goes through a vapor state. Finally, the nanoparticle-based deposition system (NPDS) consists of direct deposition of dry nanoparticulate as an aerosol onto a substrate, applying low vacuum conditions at ambient temperature [46].

Many different methods can be applied in electrochromic device construction, and the development of new techniques is frequently addressed in recent research. One of such new techniques is the maskless micropatterning of Prussian Blue nanoparticles, as described by Shimojo [20]. It is based on the focused electron beam irradiation of Prussian Blue particulate for selective fixation onto a substrate (Fig. 5).

Initially, the nanoparticulate is uniformly spread on the substrate surface. Then, during electron beam irradiation, organic molecules surrounding the nanoparticulate are decomposed into a carbonaceous substance that immobilizes the nanoparticles. Since the focused electron beam trajectory is programmable, it is possible to selectively fixate the desired micropattern. Lastly, the non-fixated portion is removed, and the final system is used in an electrochromic device.

The presented technique produced patterns with resolutions of $< 1\ \mu m$, with an electrochromic response between 100 and 120 ms. Unlike conventional lithography techniques, this method did not require complicated processes of masking, exposure, and other chemical treatments.

6 Characterization

According to which properties it is desired to be studied, an extensive range of techniques is available for electrochromic materials characterization. Usually, it is of interest the crystalline and electronic structure of the material and the ionic intercalation rate and optical properties of its films [7, 47]. Therefore, some of the techniques for the characterization of electrochromic films are presented in Table 2.

A well-known characterization technique used to study the electrochromic effect of thin films is cyclic voltammetry. For this technique, a voltage is applied to the electrochromic device in question at a known rate (V/s) while simultaneously measuring

Table.2 Characterization techniques applied for electrochromic films

Optical properties	
Ellipsometry	
Attenuated total reflectance (ATR)	
Spectrophotometry	Transmittance, Reflectance
Ion intercalation and deintercalation	
Microbalance measurements	
Nuclear magnetic resonance (NMR)	
Coulometric titration	
Chronoamperometry	
Cyclic voltammetry	
Impedance spectrometry	
Beam deflectometry	Mirage effect, Beam bending
Crystal structure and elemental composition	
Transmission electron microscopy (TEM)	
Scanning electron microscopy (SEM)	
Secondary ion mass spectroscopy (SIMS)	
Rutherford backscattering spectroscopy (RBS)	
Nuclear reaction analysis (NRA)	
X-ray fluorescence (XRF)	
X-ray extinction	Diffraction (XRD), Scattering, Absorption (EXAFS, XANES)
Vibrational spectroscopy	Infrared absorption (IR, FTIR); Raman
Electronic structure	
Electron paramagnetic resonance (EPR)	
Electron energy loss spectroscopy (EELS)	
Photoelectron spectroscopy	X-ray (XPS), Ultraviolet (UPS), Auger (AES)

the current between the working electrode and the counter electrode [48]. The potential ramp is inverted by reaching a determined voltage value, and the voltage is decreased until reaching the initial state. A cycle is completed when the voltage reaches the defined maximum value, decreases until the specified minimum negative value, and increases again until no voltage is applied. The information acquired is utilized to create a cyclic voltammogram.

This analysis not only tests the electrochromic capacity of the device but also makes it possible to observe the capability of ion intercalation/deintercalation, stability, and reaction reversibility [49]. The cyclic voltammogram of a WO_3 thin film, presented by Habib et al. [19], shows an anodic peak at the negative voltage region, characteristic of this material. It also presents the cyclic voltammogram of a Prussian Blue film. Both anodic and cathodic peaks are evident, caused by the reduction into Prussian White and posterior reoxidation into Prussian Blue (Fig. 6).

7 Applications

Since the initial studies on electrochromic materials, its application in display technology was contemplated [7, 47]. Many properties of inorganic electrochromic materials are interesting for such application, including its very low energy consumption due to the persistence of its color after ceasing voltage application. Early on, however, electrochromic technology still had some obstacles to overcome, such as low refresh rates and inferior life expectancy [50].

Eventually, the display market was dominated by liquid crystal display technology (LCD) [22], which already had the capability of switching between opaque and colored states in fractions of a second. Still, electrochromic displays present great potential in electronic paper technology.

Inorganic electrochromic devices also have important applications in electrochromic windows and mirrors, also known as “smart windows” (Fig. 7a) [14]. By adjusting the window transmittance through the electrochromic device, it is possible to block the passage of sunlight, dynamically changing the amount of thermal radiation received through it [8, 14, 22]. This allows an increase of up to 25% in the energetic efficiency of large buildings [1]. In the automobile industry, some consumer vehicles are already equipped with anti-reflective electrochromic mirrors (Fig. 7b). In addition, in some commercial planes, passengers can adjust the amount of light shining through electrochromic windows [22].

7.1 *Electrochromic Devices*

The basic configuration of a generic electrochromic device is represented in Fig. 8. This configuration is often called a thin-film battery, for its similar working to that of an electrical battery, even sharing numerous properties [14]. It consists of an active

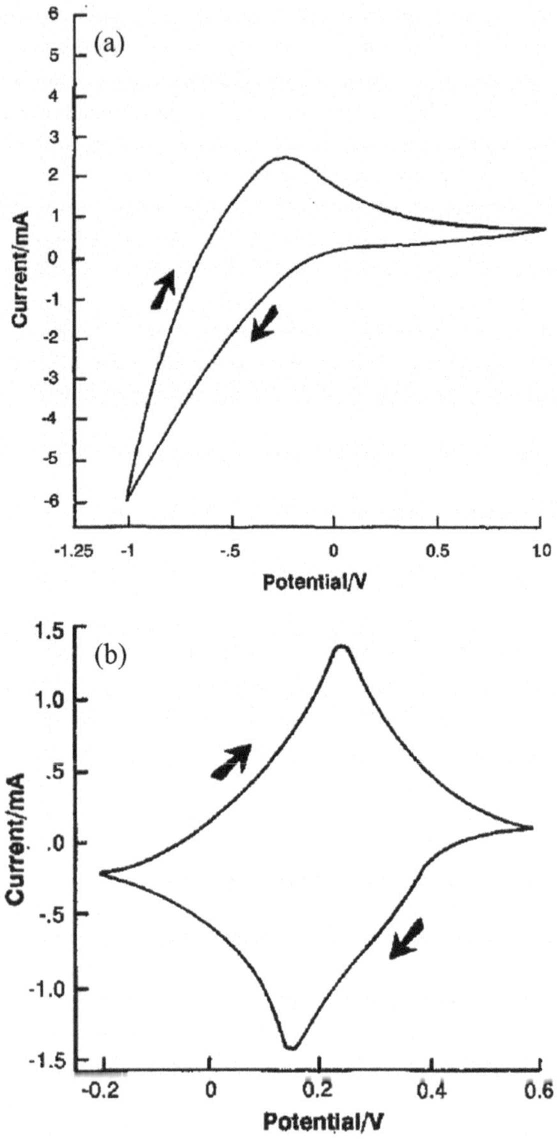

Fig. 6 **a** Cyclic voltammogram of a WO_3 film in 1 M $LiClO_4$ solution in propylene carbonate (scan rate: 100 mV/s). **b** Cyclic voltammogram of a PB-film in an aqueous solution of 1 M KCl (pH 4); scan rate: 10 mV/s. Reprinted with permission from Habib et al. [19]

electrochromic layer electrode, a counter electrode comprised of an ion storage layer, separating both electrodes an ion-conducting layer.

Both electrodes are usually fabricated on transparent substrates coated with transparent electrical conductors, commonly glass or a flexible plastic coated with indium tin oxide (ITO) [7, 14, 51]. It is through this conductive layer that the voltage will be applied to the device. One of these substrates may also not be transparent, depending on the device application, like in electrochromic mirrors. Moreover, the chosen ion

Fig. 7 Electrochromic applications: **a** dimmable and smart windows, **b** antiglare devices. Reprinted with permission from Kobayashi [22]

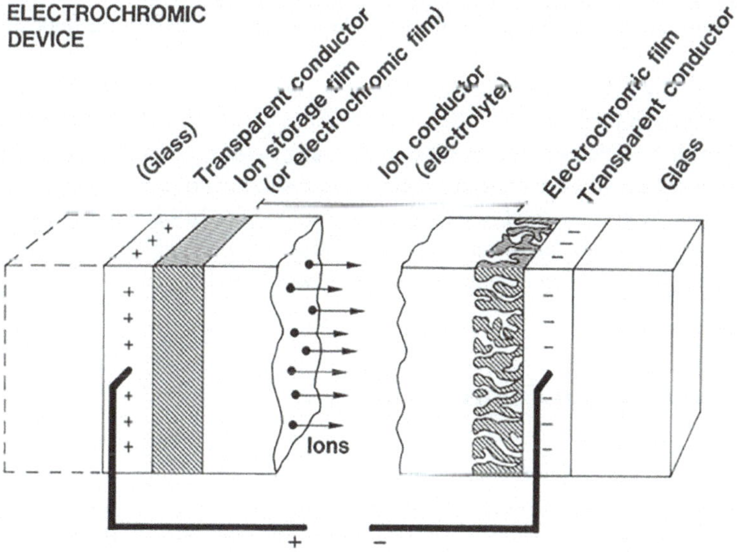

Fig. 8 Generic battery-type EC device design. Arrows indicate the movement of ions when applying voltage. Reprinted with permission from Granqvist [8]

storage layer may have complementary electrochromic properties to improve color formation [51, 52]. Finally, the ion-conducting layer may be a liquid electrolyte, a polymer, gel, or a thin film [53].

By applying a voltage to this device, the generated electric field causes the intercalation or deintercalation of ions through the ion conductor to the electrochromic layer, causing reversible color formation. To facilitate this ion migration, small ions like protons (H^+) and lithium ions (Li^+) are usually preferred [8, 14]. Additionally, the electrochromic film composition and structure significantly contribute to the overall performance of the electrochromic device. Because of that, advances in nanotechnology enable improvements in its properties, including better stability, modulation time, and better color formation [8].

During electrochromic device fabrication, deposited thin films of higher porosity enhance intercalation/deintercalation rates, which improves its performance related to color formation times. Historically, desired WO_3 films with better properties were considered porous depositions of highly disordered 'molecular solids', yet not exactly amorphous [7]. Nowadays, however, it is known that nanosized particles can cause differing electronic and physical properties through quantum size effects and present higher dispersion due to a drastic increase in surface/volume ratio [53]. Since a higher percentage of active material is concentrated at the surfaces of the particles, if well dispersed in the electrolyte, a drastic increase of the electrochromic device performance is observed.

A significant number of electrical devices and semiconductors are fabricated through deposition techniques, including electrochromic devices. As a result, deposition techniques are constantly being improved and developed [46]. One of these techniques, based on physical vapor deposition, enables high precision nano sculpting of columnar thin films. Oblique angle deposition (OAD) consists of the deposition of films through PVD while maintaining the substrate at a particular angle concerning the particle flux, resulting in nucleation sites in some parts of the substrate and columnar growth [16].

Furthermore, it is possible to have even greater control of the film final structure by precisely and gradually rotating the substrate, further evolving the OAD technique. This new method, called glancing angle deposition (GLAD), produces distinct and precise surface structures with distinct optical and electrical properties [16], increasing the potential applications of these films (Fig. 9).

8 Conclusion and Future Perspective

This work introduced electrochromic materials and discussed inorganic and organic-based electrochromic characterization techniques, basic device construction, and application. As it was shown, various kinds of materials may present electrochromism. In addition, advances in nanotechnology enabled the development of new and better performing electrochromic materials and devices [53].

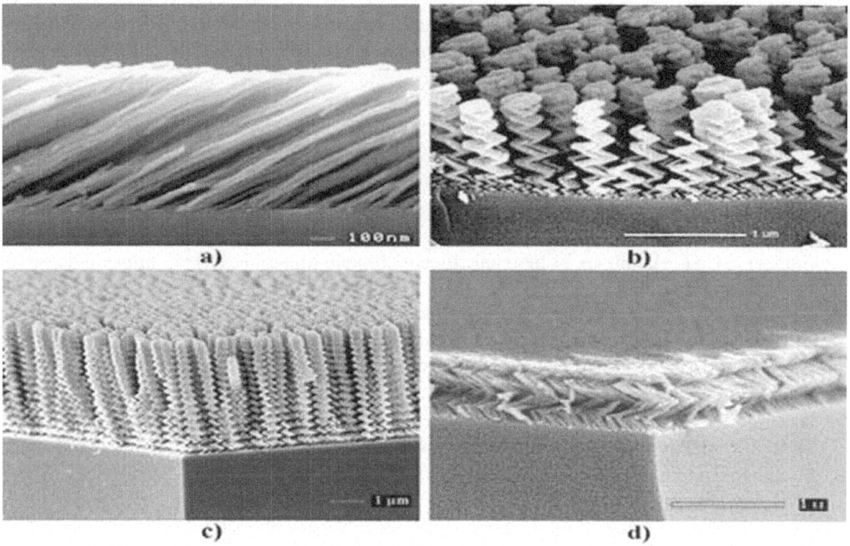

Fig. 9 Nanostructured MgF_2 films deposited using GLAD, at static incidence flux angle of 85°, under different deposition conditions. Reprinted with permission from Gupta et al. [16]

In terms of applications, electrochromic devices such as smart windows have usually been employed in the niche or high-end commercial products and have yet to enter a wider market as a commodity product. However, this situation seems to be likely changing due to the increased need for environmentally friendly technologies [54]. Electrochromics is considered one of the "green nanotechnologies" with increasing interest in recent times [8].

Publications in electrochromics are increasing sharply [23, 53], which indicates the growing potential of this technology. In addition, new electrochromic devices for multispectral energy modulation have also been reported [25], including radiation in the near-infrared, thermal infrared, and microwave regions. These could, for example, significantly extend the modulation capacity of solar radiation of smart windows through multi-functional devices [54].

References

1. Lampert, C.M.: Electrochromic materials and devices for energy efficient windows. Solar Energy Mater. **11**, 1–27 (1984). https://doi.org/10.1016/0165-1633(84)90024-8
2. Evecan, D., Zayim, E.: Highly uniform electrochromic tungsten oxide thin films deposited by e-beam evaporation for energy saving systems. Curr. Appl. Phys. **19**, 198–203 (2019). https://doi.org/10.1016/j.cap.2018.12.006
3. Granqvist, C.G.: Electrochromics for smart windows: oxide-based thin films and devices. Thin Solid Films **564**, 1–38 (2014). https://doi.org/10.1016/j.tsf.2014.02.002

4. Granqvist, C.G., et al.: Electrochromic materials and devices for energy efficiency and human comfort in buildings: a critical review. Electrochim. Acta **259**, 1170–1182 (2018). https://doi.org/10.1016/j.electacta.2017.11.169
5. Pittaluga, M.: Electrochromic glazing and walls for reducing building cooling needs. In: Eco-Efficient Materials For Mitigating Building Cooling Needs: Design, Properties and Applications (2015). http://doi.org/https://doi.org/10.1016/B978-1-78242-380-5.00017-0
6. Wu, C.-M., Naseem, S., Chou, M.-H., Wang, J.-H., Jian, Y.-Q.: Recent advances in tungsten-oxide-based materials and their applications. Front. Mater. **6**, 49 (2019). https://doi.org/10.3389/fmats.2019.00049
7. Granqvist, C.G.: Handbook of Inorganic Electrochromic Materials. (1995). https://doi.org/10.1016/B978-0-444-89930-9.X5000-4
8. Granqvist, C.G.: Switchable glazing technology: electrochromic fenestration for energy-efficient buildings. In: Pacheco Torgal, F., Mistretta, M., Kaklauskas, A., Granqvist, C., Cabeza, L. (eds.) Nearly Zero Energy Building Refurbishment. Springer, London (2013). https://doi.org/10.1007/978-1-4471-5523-2_21
9. Somani, P.R., Radhakrishnan, S.: Electrochromic materials and devices: present and future. Mater. Chem. Phys. **77**, 117–133 (2003). https://doi.org/10.1016/S0254-0584(01)00575-2
10. Mortimer, R.J., Rosseinsky, D.R., Monk, P.M.S.: Electrochromic Materials and Devices. Wiley-VCH, Weinheim, Germany (2015)
11. Tarver, J., Yoo, J.E., Loo, Y.L.: Organic electronic devices with water-dispersible conducting polymers. Comprehensive Nanosci. Technol. **5**, 413–446 (2011). https://doi.org/10.1016/B978-0-12-374396-1.00136-7
12. Granqvist, C.G., Pehlivan, I.B., Niklasson, G.A.: Electrochromics on a roll: web-coating and lamination for smart windows. Surface Coat. Technol. **336**,133–138. https://doi.org/10.1016/j.surfcoat.2017.08.006
13. Peng, H., Fang, X.: Electrically driving sensors based on polymer. In: Peng, H., Weng, W., Sun, X., Fang, X. (eds.) Polymer Materials for Energy and Electronic Applications. Academic Press, London (2017). https://doi.org/10.1016/C2015-0-01541-6
14. Tan, X.Y., Wang, H., Kim, T.G.: electrochromic smart windows: an energy-efficient technology. In: Siddiquee, S., Gan Jet Hong, M., Mizanur Rahman, M. (eds.) Composite Materials: Applications in Engineering, Biomedicine and Food Science. Springer, Cham (2020). https://doi.org/10.1007/978-3-030-45489-0_9
15. Weil, M., Schubert, W.-D.: The Beautiful Colours of Tungsten Oxides. Tungsten Newsletter, International Tungsten Industry Association (2013)
16. Gupta, J., Shaik, H., Kumar, K.N.: A review on the prominence of porosity in tungsten oxide thin films for electrochromism. Ionics **27**, 2307–2334 (2021). https://doi.org/10.1007/s11581-021-04035-8
17. Li, C., Hsieh, J.H., Humg, M.-T., Huang, B.Q.: Electrochromic study on amorphous tungsten oxide films by sputtering. Thin Solid Films **587**, 75–82 (2014). https://doi.org/10.1016/j.tsf.2014.12.022
18. Rowley, N.M., Mortimer, R.J.: New electrochromic materials. PubMed **85**, 243–262 (2020) https://doi.org/10.3184/003685002783238816
19. Habib, M.A., Maheswari, S.P.: Electrochromic characteristics of a complementary tungsten trioxide/Prussian blue cell. J. Appl. Electrochem. **23**, 44–50 (1993). https://doi.org/10.1007/BF00241574
20. Shimojo, M., Suzaka, Y., Katori, Y.: Maskless micropatterning of electrochromic nanoparticles using a focused electron beam. J. Vac. Sci. Technol. B **37** (2019). https://doi.org/10.1116/1.5080754
21. Welsh, T.A., Draper, E.R.: Water soluble organic electrochromic materials. RSC Adv. **11**, 5245–5264 (2021). https://doi.org/10.1039/D0RA10346B
22. Kobayashi, N.: Electrochromic displays. In: Chen, J., Cranton, W., Fihn, M. (eds.) Handbook of Visual Display Technology. Springer, Berlin, Heidelberg (2015). https://doi.org/10.1007/978-3-642-35947-7_188-1

23. Oliveira, R.S., Oliveira, M.R., Oliveira, S., Ponzio, E.A.: Organic electrochromic materials: a brief review of viologens, phthalocyanines and some complexes of transition metals. Revista Virtual de Química **5**, 596–629 (2013). https://doi.org/10.5935/1984-6835.20130044
24. Mortimer, R.J.: Organic electrochromic materials. Electrochim. Acta **44**, 2971–2981 (1999). https://doi.org/10.1016/S0013-4686(99)00046-8
25. Zmija, J., Malachowski, M.J.: New organic electrochromic materials and their applications. J. Achiev. Mater. Manuf. Eng. **48**, 14–23 (2011)
26. Michaelis, L., Hill, E.S.: The viologen indicators. J. Gen. Physiol. **16**(6), 859–873 (1933). https://doi.org/10.1085/jgp.16.6.859
27. Summers, L.A.: The Bipyridinium Herbicides. Academic, New York (1980)
28. Inzelt, G.: In: Bard, A.J. (ed.) Electroanalytical Chemistry, vol. 18. Marcel Dekker, New York (1994)
29. Ostrom, G.S., Buttry, D.A.: J. Phys. Chem. **99**(15), 236 (1995)
30. Bouridah, A., Dalard, F., Armand, M.: Solid State Ionics **28–30**, 950 (1988)
31. Bird, C.L., Kuhn, A.T.: Chem. Soc. Rev. **10**, 49 (1981)
32. Hwang, E., et al.: an electrolyte-free flexible electrochromic device using electrostatically strong graphene quantum dot-viologen. Adv. Mater. **26**, 5129–5134 (2014). https://doi.org/10.1002/adma.201401201
33. Claessens, C., Blau, W., Cook, M., et al.: Phthalocyanines and phthalocyanine analogues: the quest for applicable optical properties. Monatsh. Chem. **132**, 3–11 (2001). https://doi.org/10.1007/s007060170140
34. Madhuri, K.P., Kaur, P., Ali, M.E., John, N.S.: Nanoscale conductance in lead phthalocyanine thin films: influence of molecular packing and humidity. J. Phys. Chem. C **121**, 9249–9259 (2017). https://doi.org/10.1021%2Facs.jpcc.6b09240
35. Mori, S., Nagata, M., Nakahata, Y., et al.: Enhancement of incident photon-to-current conversion efficiency for phthalocyanine-sensitized solar cells by 3D molecular structuralization. J. Am. Chem. Soc. **132**, 4054–4055 (2010). https://doi.org/10.1021/ja9109677
36. Saito, Y., Higuchi, T., Sugimori, H., Yabu, H.: One-Pot UV-assisted synthesis of metal phthalocyanine nanocrystals. Chem. Nano. Mat. **1**, 92–95 (2015). https://doi.org/10.1002/cnma.201500017
37. Mu, J., Shao, C., Guo, Z., Zhang, M., Zhang, Z., Zhang, P., et al.: Solvothermal synthesis and electrochemical properties of 3D flower-like iron phthalocyanine hierarchical nanostructure. Nanoscale **3**, 5126–5131 (2011). https://doi.org/10.1039/C1NR10627A
38. Tong, W.Y., Djurisic, A.B., Xie, M.H., et al.: Metal Phthalocyanine Nanoribbons and Nanowires. J. Phys. Chem. B **110**, 17406–17413 (2006). https://doi.org/10.1021/jp062951q
39. Madhuri, K.P., John, N.S.: Metallophthalocyanine-nanofibre-based electrodes for electrochemical sensing of biomolecules. Bull. Mater. Sci. **41**, 118 (2018). https://doi.org/10.1007/s12034-018-1636-9
40. Banasz, R., Walesa-Chorab, M.: Polymeric complexes of transition metal ions as electrochromic materials: Synthesis and properties. Coordination Chem. Rev. **389**, 1–18 (2019). https://doi.org/10.1016/j.ccr.2019.03.009
41. Schott, M., Beck, M., Winkler, F., et al.: Fabricating electrochromic thin films based on metallopolymers using layer-by-layer self-assembly: an attractive laboratory experiment. J. Chem. Educ. **92**, 364–367 (2015). https://doi.org/10.1021/ed5002174
42. Higuchi, M.: Stimuli-responsive metallo-supramolecular polymer films: design, synthesis and device fabrication. J. Mater. Chem. C **2**, 9331–9341 (2014). https://doi.org/10.1039/C4TC00689E
43. Han, F.S., Higuchi, M., Kurth, D.G.: Metallosupramolecular polyelectrolytes self-assembled from various pyridine ring-substituted bisterpyridines and metal ions: photophysical, electrochemical, and electrochromic properties. J. Am. Chem. Soc. **130**, 2073–2081 (2008). https://doi.org/10.1021/ja710380a
44. Roy, S., Chakraborty, C.: Sub-second electrochromic switching and ultra-high coloration efficiency in halloysite nanoclay incorporated metallo-supramolecular polymer nano-hybrid based electrochromic device. Sol. Energy Mater. Sol. Cells **208**, 110392 (2020). https://doi.org/10.1016/j.solmat.2019.110392

45. Niklaus, L., Schott, M., Subel, J., et al.: Mixed metal oxides as optically passive ion storage layers in electrochromic devices based on metallopolymers. Sol. Energy Mater. Sol. Cells **223**, 110950 (2021). https://doi.org/10.1016/j.solmat.2020.110950
46. Park, S.-I., Quan, Y.J., Kim, S.H. et al.: A review on fabrication processes for electrochromic devices. Int. J. Precis. Eng. Manuf.-Green Tech. **3**, 397–421 (2016). https://doi.org/10.1007/s40684-016-0049-8
47. Faughnan, B.W., Crandall, R.S.: Electrochromic displays based on WO3. In: Pankove, J.I. (ed.) Display Devices. Topics in Applied Physics, vol 40. Springer, Berlin, Heidelberg (1980). https://doi.org/10.1007/3540098682_12
48. Bilal, S.: Cyclic voltammetry. In: Kreysa, G., Ota, K., Savinell, R.F. (eds.) Encyclopedia of Applied Electrochemistry. Springer, New York, NY (2014). https://doi.org/10.1007/978-1-4419-6996-5_220
49. Ristova, M.M., Mirceski, V., Neskovska, R.: Voltammetry of chemically deposited Cu x O electrochromic films, coated with ZnO or TiO2 electrocatalyst layers. J. Solid State Electrochem. **19**, 749–756 (2015). https://doi.org/10.1007/s10008-014-2666-x
50. Tarver, J., Yoo, J.E., Loo, Y.-L.: Organic electronic devices with water-dispersible conducting polymers. In: Andrews, D.L., Wiederrecht, G.P., Scholes, G.D. (eds.) Comprehensive Nanoscience and Technology, pp. 413–446 (2011). https://doi.org/10.1016/B978-0-12-374396-1.00136-7
51. Heusing, S., Aegerter, M.A.: Sol-Gel coatings for electrochromic devices. In: Aparicio, M., Jitianu, A., Klein, L. (eds.) Sol-Gel Processing for Conventional and Alternative Energy. Advances in Sol-Gel Derived Materials and Technologies. Springer, Boston, MA (2012). https://doi.org/10.1007/978-1-4614-1957-0_12
52. Habib, M.A., Maheswari, S.P., Carpenter, M.K.: A tungsten-trioxide/prussian blue complementary electrochromic cell with a polymer electrolyte. J. Appl. Electrochem. **21**, 203–207 (1991). https://doi.org/10.1007/BF01052571
53. Vidotti, M., Torresi, S.I.C.: Nanochromics: old materials, new structures and architectures for high performance devices. J. Braz. Chem. Soc. **19** (2008). https://doi.org/10.1590/S0103-50532008000700004
54. Granqvist, C.G., Bayrak Pehlivan, İ., Green, S.V., et al.: Oxide-based electrochromics: advances in materials and devices. MRS Online Proc. Lib. **1328**, 201 (2011). https://doi.org/10.1557/opl.2011.1064

Synthesis and Characterization of Nb_2O_5 Nanostructures

Thais Cristina Lemes Ruwer

Abstract In the last few years, there has been a significant increase in the interest in niobium oxides, among which Nb_2O_5 stands out. This interest is due to its enormous potential in several applications such as microalloys, thin films, medical implants, superalloys, superconductors, electrolytics, supercapacitors, electronics, batteries, electrodes, solar cells, among others. This chapter will address the different forms for synthesizing Nb_2O_5 nanostructure and the main characteristics and applications.

Keywords Nb_2O_5 · Nanostructure · Nanomaterials · Nanotechnology

1 Introduction

Brazil currently has more than 90% of the world's exploitable niobium reserves. It is also the largest producer and exporter of this metal. The search for new nanomaterials, including niobium nanomaterials, is extremely important, as the knowledge of these materials presents an evolution in terms of discoveries of physicochemical properties, opening paths for new applications.

In the last 20 years, there has been a considerable increase in investment to develop nanotechnology to produce nanomaterials. As a result, the nanomaterials market may reach US$ 100 billion in 2025 [1]. Nanomaterials offer numerous features due to their specific physical and chemical properties, which favor their use to meet the needs of several applications [2].

Nanotechnology offers new paths that open an opportunity to supply needs that materials in the bulk form previously did not meet satisfactorily. Nanotechnology also finds applications in the sectors of food and agriculture, consumer products, transport and logistics, energy, and environmental sectors. The number and variety of new and modified nanomaterials developing and the range of applications have increased significantly over the past ten years [3].

Due to the interesting properties, the literature presents numerous applications of Nb_2O_5, such as solid electrolytic capacitors, catalysis, photochromic devices,

T. C. L. Ruwer (✉)
Universidade Federal Do Rio Grande Do Sul, Porto Alegre, Brazil

A. Kopp Alves (ed.), *Technological Applications of Nanomaterials*,
Engineering Materials, https://doi.org/10.1007/978-3-030-86901-4_9

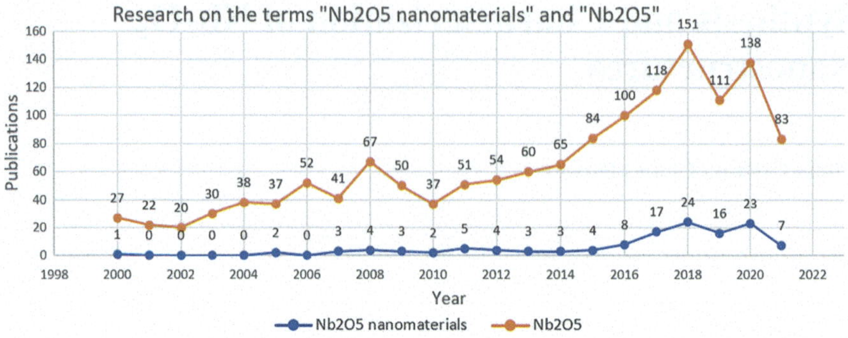

Fig. 1 The number of worldwide publications on "Nb_2O_5 nanomaterials" between the years 2000 and 2020

conductive oxides; that is, it is essential and urgent to understand the physical properties of niobium oxides and your control.

In Fig. 1, the graph shows the number of worldwide publications on "Nb_2O_5 nanomaterials" between the years 2000 and 2020. We can observe a significant increase in publications related to "Nb_2O_5", showing that this material is very researched. However, the term "Nb_2O_5 nanomaterials", although there is an increase, is much less expressive; that is, there is an unknown path about Nb_2O_5 nanomaterials, there is still a lot to be researched [4].

This chapter provides an overview of methods for obtaining Nb_2O_5 nanostructures, characterization, and applications.

2 Niobium Pentoxide (Nb_2O_5)

Niobium pentoxide (Nb_2O_5) is an essential n-type semiconductor with a bandgap of around 3.4 eV; its remarkable physicochemical properties and structure make it suitable for various applications in various areas such as catalysis, sensors, health, aerospace, among others.

Nb_2O_5 is usually presented as a white solid, stable in air and insoluble in water; it can be dissolved in both strong acids and strong bases. Although Nb_2O_5 generally has a crystalline structure of NbO_6 octahedrons, this structure can be distorted to different degrees. In some phases, we can find different structures for coordinating polyhedra, such as NbO_7 and NbO_8 [4, 5].

Nb_2O_5 can occur in the amorphous state or as crystalline polymorphs. Polymorphic Nb_2O_5 is usually presented in white powder form, and monocrystalline Nb_2O_5 is transparent. The physical properties of Nb_2O_5 depend on its polymorphism and the technique of the chosen synthesis parameters [6].

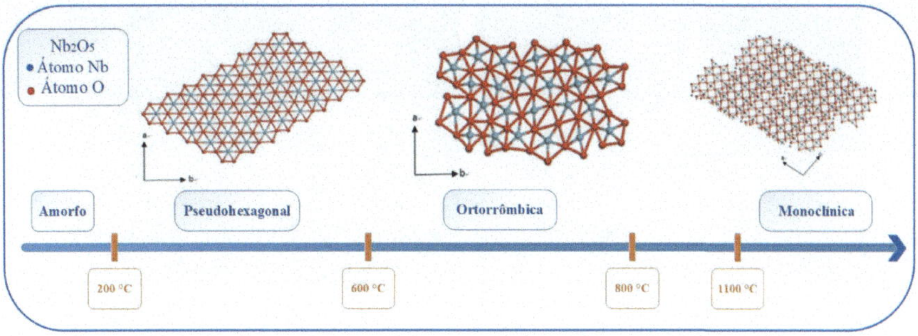

Fig. 2 Evolution of the crystalline phases of Nb_2O_5 as a function of temperature. Edited and reprinted with permission the from Springer [10]

Nb_2O_5 has high structural complexity due to the characteristic polymorphism, whose degree is related to the method and variables of the synthesis, such as the nature of the precursors, processing time, and temperature.

The literature does not agree with the definitions and terminology of crystalline phases of Nb_2O_5, the main phases reported are pseudohexagonal (TT- Nb_2O_5) for treatments with low temperatures (up to 600 °C), orthorhombic structure (T-Nb_2O_5) for treatments using temperatures from 600 °C to 800 °C, and monoclinic structure (H-Nb_2O_5) for the use of temperatures above 1100 °C, as shown in Fig. 2 [7, 89].

3 Synthesis Methods

There are numerous forms and processes usually for obtaining Nb_2O_5. The selection of the synthesis method is due to its ability and practicality of regulating Nb_2O_5 properties. Here I will present the main synthesis methods most used.

3.1 Anodizing Process

The anodizing process is the most researched and used method of obtaining nanostructures globally, as it is a very efficient method in forming highly ordered, porous, and symmetrical oxides. In addition, Anodizing is a simple method, with relatively low cost and a high level of reproducibility [11].

The anodizing process's structure consists of two electrodes being a working electrode and a counter electrode immersed in a liquid electrolyte and then applying a potential difference between them. This applied potential difference can be in voltage or electric current, generating electrochemical reactions on the working electrode surface. Consequently, an oxide film is formed on the metal surface. The morphology and growth of the oxide formed directly influence the applied anodizing potential, the

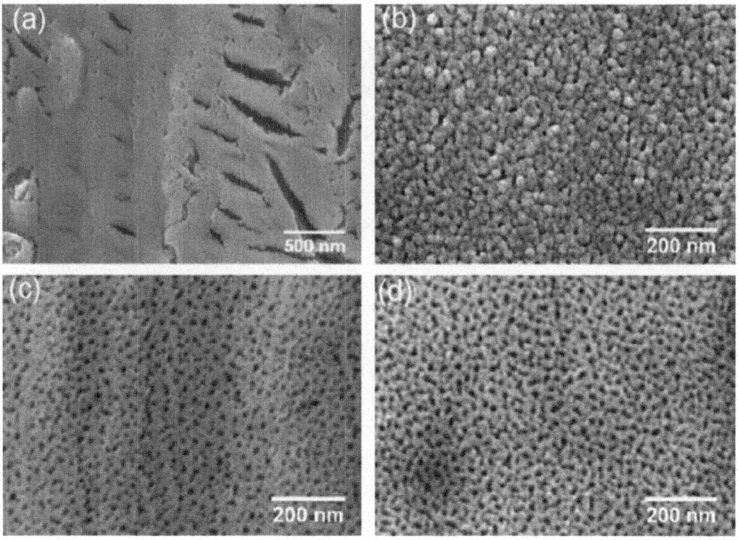

Fig. 3 Scanning electron micrographs of the surfaces of niobium specimens anodized with glycerol electrolytes with **a** 0.08 **c** 0.16 and **d** 0.65 mass % water at 433 K. The surface of the inner oxide layer after removal of the outer oxide layer formed in the electrolyte with 0.08 mass% water is shown in (**b**). Reprinted with permission from [13]

electrolyte composition, the electrolyte pH, the electrolyte temperature, the agitation used during the process, and the anodizing time [9, 10].

Nb_2O_5 usually synthesized by anodization is amorphous. That is, they need a thermal treatment for the crystallization of Nb_2O_5 to occur. Draper presented research on the structure of anodic films formed from niobium to understand how the film's material can influence the growth rate [12].

Habazaki demonstrated the importance of water content in the formation of porous anodic niobium oxide films in hot phosphate-glycerol electrolytes. He concluded that water in the electrolyte is a predominant source of oxygen in the formed structure. The Nb_2O_5 films formed in the electrolyte containing 0.65% by weight of water were almost free of phosphorus species. The reduction in water content increased the incorporation of phosphorus species, as shown in Fig. 3 [13].

3.2 *Sol–gel Process*

The sol–gel definition is used to present a variety of material synthesis processes in which a solid phase (gel) is formed through the gelfication of a colloidal suspension (sol). This method is widely used for the synthesis of niobium oxides, especially Nb_2O_5. After obtaining this gel, a thermal treatment is usually carried out to remove the residues from the synthesis, stabilize the gel and crystallize it [14].

Fig. 4 Sample **a** wet gel and **b** dried gel. Reprinted with permission from Springer [15]

According to Aronne, the flexibility of the sol–gel chemistry allows for greater control over the form, morphology, structure, and properties of the material obtained through the process parameters (pH, concentration, molar ratio, and temperature). For example, Aronne used the sol–gel process to synthesize transparent chemical gels from mixed niobium-silicon oxide at room temperature. Figure 4 shows the wet and dry gels [15].

The Nb_2O_5-SiO_2 gels were synthesized from niobium chloride and tetraethoxysilane under acidic conditions, the role of HCl was presented as a niobium complexing agent. He concluded that the degree of dispersion of niobium is very high, even if strongly influenced by the Nb/Si ratio; that is, this proportion influences both the tendency of crystallization and the coordination of niobium in the matrix [15].

3.3 *Hydrothermal Process*

Li Qin presented in his research a straightforward and efficient hydrothermal process, from multi-layered niobium a compound named Nb_2CT_x/Nb_2O_5 was obtained as shown in Fig. 5. This hybrid material resulting from Nb_2O_5 showed excellent characteristics when used as anodes for lithium-ion batteries. The Nb_2O_5 nano-rods were grown in situ on the surfaces of the material obtained from multilayer Nb_2CT_x by means of a single-stage hydrothermal process, thus Nb_2CT_x acted as a source of niobium and as a conducting medium simultaneously [16].

Based on this strategy, the 2D structure of Nb_2CT_x was stabilized in parallel, generating a hierarchical hybrid Nb_2CT_x/Nb_2O_5. Nb_2CT_x as a conductive matrix

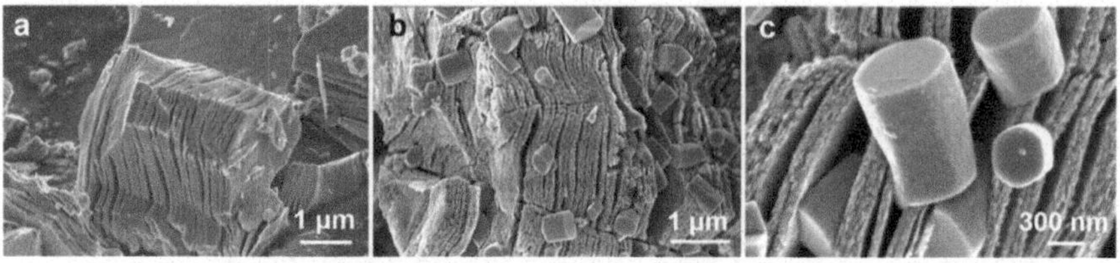

Fig. 5 **a** FESEM images of Nb_2CTx. **b** and **c** FESEM images of Nb_2CTx/Nb_2O_5. Reprinted with permission from [16]

compensated for the insufficient electrical conductivity of Nb_2O_5, and the 2D structure retained was favorable for the rapid transport of Li^+. The hybrid material obtained Nb_2CT_x/Nb_2O_5 showed excellent performance of stable reversible capacity, rate, and long life cycle, highlighting its enormous potential in the application in high-performance lithium-ion batteries [16].

Hao Wen synthesized nanorods from niobium foil using a straightforward hydrothermal process. As a result, nb_2O_5 nanorods with approximately 100 nm in diameter and 1 μm in length were obtained. The characterizations indicated that an orthorhombic structure was obtained that grew longitudinally along the direction (0 0 1). NH_4F was used as a precursor that, together with the metal plate, was placed in a Teflon-coated autoclave and kept at 150 °C in the oven for 48 h. After this process, the plate was dried at 50 °C in the air for 12 h. Finally, annealing was carried out at 300–500 °C for 30 min in the air. Figure 6 shows the nanostructure formed [9].

There was no change in the nanostructure microstructure, carrying out the treatment at temperatures below 500 °C; however, the crystallinity improved. The material obtained from Nb_2O_5 was tested as a cathodic material for lithium batteries, and it showed a good specific capacity of up to 380 $mAhg^{-1}$ even after 50 charges/discharge cycles [9].

Fig. 6 SEM images of samples synthesized at 150 °C for 48 h in aqueous NH_4F solution with a concentration of 0.02 M. Reprinted with permission from [9]

3.4 Microwave-Assisted Hydrothermal Process

Wermuth synthesized $KNbO_3$ perovskite nanostructures from Nb_2O_5 and KOH via microwave-assisted hydrothermal synthesis to degrade rhodamine B. The results showed that perovskite with molar ratio [1:16] showed better photocatalytic efficiency in wheel degradation -mina B. The large specific surface area and more efficient conversion of UV photons into effective charge carriers could explain its better performance demonstrating the excellent applicability potential of this nanostructured $KNbO_3$ material in the degradation of organic pollutants in wastewater. Figure 7 shows the nanostructure obtained from $KNbO_3$ [17].

Ruwer synthesized homogeneous Nb_2O_5 nanostructures very quickly using a microwave-assisted hydrothermal method. Metallic niobium plates were used to form the nanostructure. Microwave irradiation reduced the synthesis time from several days to 2 h. Figure 8 shows the formation of nanorod matrices structured at 200 °C [18].

The results indicated that the microwave-assisted hydrothermal method is a fast and inexpensive technique for producing homogeneous matrices of Nb_2O_5 nanorods [18].

Fig. 7 SEM images of the $KNbO_3$ synthesized in the molar ratios **a** [1:8], **b** [1:12], and **c** [1:16] for 30 min of microwave-assisted hydrothermal synthesis. Reprinted with permission from [17]

Fig. 8 SEM images of the sample treated in the microwave at 200 °C for 2 h. Reprinted with permission from [18]

4 Characterization Methods

4.1 Characterization by DRX

Braga synthesized copper oxide and niobium pentoxide supported on silica-alumina by the co-impregnation method and by the sequential impregnation method in the following proportions: 1:10 and 1: 1 CuO: Nb_2O_5. The materials synthesized by sequential impregnation showed less thermal stability than the analogs prepared by co-impregnation based on Nb_2O_5. That is, the synthesized materials presented different characteristics according to the method of preparation. Figure 9 shows the XRD analysis with the appropriate patterns of Cu and Nb oxides and the presence of a third component (copper and niobium oxide) in the samples with CuO content $\geq$ 10% by weight when calcined at 800 °C by 6 h [19].

With heating treatment, niobium oxalate decomposes gradually into niobium pentoxide. Oxalate calcined at 700 °C for 3 h presented a pattern similar to niobium pentoxide treated at 800 °C for 6 h, with a predominantly orthorhombic phase. At heating temperatures of about 300–400 °C for 6 h, the pure crystalline oxalate showed a halo-amorphous pattern, similar to the pattern observed for Nb_2O_5/SiO_2-Al_2O_3 in heat treatment at 300 °C for 6 h. Copper nitrate degrades and forms copper oxide at 300 °C, exhibiting the most intense reflections in copper oxide and niobium pentoxide supported on silica-alumina calcined at 300 °C for 6 h showed only the two reflections about CuO, validating that this is the only crystalline phase with these treatment conditions [19].

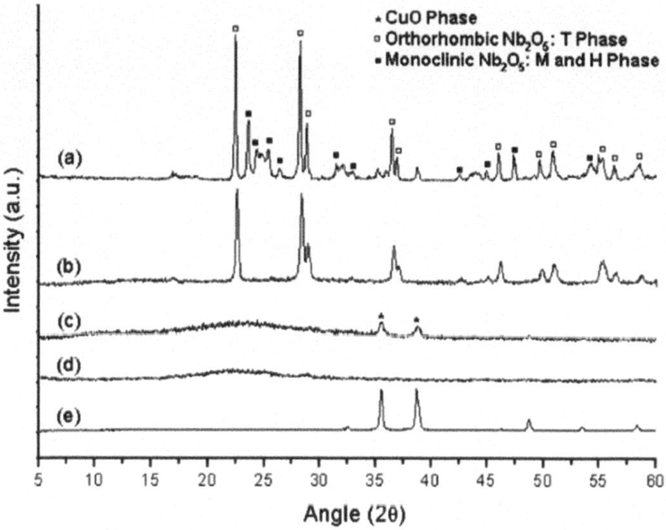

Fig. 9 XRD patterns at different calcination conditions for: Nb_2O_5 at CuO/10% Nb_2O_5/SiO_2-Al_2O_3 at 300 °C/6 h, series A (**c**); 10% Nb_2O_5/ 800 °C/6 h **a**; $NH_4[NbO(C_2O_4)_2(H_2O)_2](H_2O)_n$ at 700 °C/3 h **b**; 10% SiO_2-Al_2O_3 at 300 °C/6 h **d**; $Cu(NO_3)\cdot 3H_2O$ at 300 °C/6 h (**e**). Reprinted with permission from [19]

4.2 Characterization by Image

One of the main ways to characterize a nanostructured material is through images with equipment capable of obtaining clear images of very small structures, such as SEM. Some works will be presented that used the resource of images to present the nanostructures of different morphologies obtained from Nb_2O_5 [6].

Liu presents in his work the manufacture of Nb_2O_5 nano matrices with controlled degrees of branching. The structure was produced based on a solution with HF acid and niobium leaf as the source materials. Modifying the pH value of the reaction, the size, shape, and degree of branching of the formed structure were controlled. The property of the nanostructure can be influenced by its morphology. The growth method showed good potential as sensors and photocatalysts [20].

Figure 10a shows the SEM characterization of nano-formed sheets synthesized by a simple hydrothermal process of Nb_2O_5 that were highly crystalline with thicknesses of 3–5 nm. From cyclical medications, the material used as an electrode showed a high reversible charge/discharge capacity and cycle stability, making it a potential candidate as a positive electrode material for lithium-ion batteries [21].

Figure 10b shows nanostructures synthesized from the surface of the metallic niobium sheet using a hydrothermal process. The nanostructure was produced from the thin sheet of metallic niobium and an aqueous solution of ammonium fluoride (NH_4F), treating hydrothermally at 150 °C for 48 h. The Nb_2O_5 nanorods obtained

Fig. 10 Scanning electron microscopy (SEM) images show the morphologies of the hydrothermally synthesized Nb_2O_5. **a** Nb_2O_5-nanosheets synthesized from NbO_2 precursor at 130 °C for 30 days, **b** Nb_2O_5-nanorods synthesized at 150 °C for 48 h in NH_4F aqueous solution of 0.02 M concentration. Reprinted with permission from (**a**) [21] and (**b**) [22]

were tested as a cathodic material for lithium batteries and showed a good specific capacity [22].

Li presented the synthesis of monodispersed niobium oxide spheres from the antisolvent precipitation process shown to act as highly effective solid acid catalysts. The spherical monodispersed catalysts with size in the submicrometric range showed an essential performance in a wide range of acid-catalyzed reactions, such as Friedel–Crafts alkylation, esterification, and hydrolysis of acetates. Also, they showed good recovery capacity [23].

5 Conclusions

Although there are highly relevant works on Nb_2O_5 nanostructures, there is still a lot to explore on this material. Nanotechnology paves the way for new morphologies and applications of materials previously limited in their bulk form. Synthesizing processes are becoming easier and faster. And this is mainly due to the increase in knowledge of its remarkable physical–chemical properties and its nanostructures making it suitable for a wide range of applications.

References

1. Allan, J et al.: Regulatory landscape of nanotechnology and nanoplastics from a global perspective. Regul. Toxicol. Pharmacol. **122**(October 2020), 104885 (2021). https://doi.org/10.1016/j.yrtph.2021.
2. Gajanan, K., Tijare, S.N.: Applications of nanomaterials. Mater. Today Proc. **5**(1), 1093–1096 (2018). https://doi.org/10.1016/j.matpr.2017.11.187
3. Kraegeloh, A., Suarez-Merino, B., Sluijters, T., Micheletti, C.: Implementation of safe-by-design for nanomaterial development and safe innovation: Why we need a comprehensive approach. Nanomaterials **8**(4) (2018). https://doi.org/10.3390/nano8040239
4. Science direct. [Online] Available: www.sciencedirect.com. Accessed 28 Apr 202
5. Nico, C., et al.: Sintered NbO powders for electronic device applications. J. Phys. Chem. C **115**(11), 4879–4886 (2011). https://doi.org/10.1021/jp110672u
6. Rani, R.A., Zoolfakar, A.S., O'Mullane, A.P., Austin, M.W., Kalantar-Zadeh, K.: Thin films and nanostructures of niobium pentoxide: Fundamental properties, synthesis methods and applications. J. Mater. Chem. A **2**(38), 15683–15703 (2014). https://doi.org/10.1039/C4TA02561J
7. Prof, B.Y., Schafer, I., Gruehn, R., Schulte, F., Universitat, A.I.D.E.R., Germany, M.: The modifications of niobium pentoxide. Angew. Chem. interrnat. Edit. 1. 5(1), 40–52 (1966)
8. Raba, A.M., Bautista-Ruíz, J., Joya, M.R.: Synthesis and structural properties of niobium pentoxide powders: A comparative study of the growth process. Mater. Res. **19**(6), 1381–1387 (2016). https://doi.org/10.1590/1980-5373-MR-2015-0733
9. Wen, H., Liu, Z., Wang, J., Yang, Q., Li, Y., Yu, J.: Facile synthesis of Nb2O5nanorod array films and their electrochemical properties. Appl. Surf. Sci. **257**(23), 10084–10088 (2011). https://doi.org/10.1016/j.apsusc.2011.07.001
10. Shen, P. et al.: Nanoscale Niobium Oxides Anode for Electrochemical Lithium and Sodium Storage : A Review of Recent Improvements, vol. 0123456789. Springer Berlin Heidelberg (2020). https://doi.org/10.1007/s40097-020-00367-5
11. Ruwer, T.: Revealing the True Impact of Interstitial and Substitutional Nitrogen Doping in TiO_2 on Photoelectrochemical Applications (2021) https://doi.org/10.1039/D0TA11494D
12. Draper, P.H.G.: The structure an electron diffraction examination of the products of anodic oxidation on tantalum and zirconium. Acta Metallurgica **11** (1963)
13. Habazaki, H., et al.: Importance of water content in formation of porous anodic niobium oxide films in hot phosphate-glycerol electrolyte. Electrochim. Acta **54**(3), 946–951 (2009). https://doi.org/10.1016/j.electacta.2008.08.031
14. Lopes, O.F., De Mendonça, V.R., Silva, F.B.F., Paris, E.C., Ribeiro, C.: Óxidos de nióbio: Uma visão sobre a síntese do Nb_2O_5e sua aplicação em fotocatálise heterogênea. Quim. Nova **38**(1), 106–117 (2015). https://doi.org/10.5935/0100-4042.20140280
15. Aronne, A., et al.: Sol-gel synthesis and structural characterization of niobium-silicon mixed-oxide nanocomposites. J. Sol-Gel Sci. Technol. **43**(2), 193–204 (2007). https://doi.org/10.1007/s10971-007-1563-5
16. Qin, L., Xu, S., Liu, Y., Zhu, S., Hou, L., Yuan, C.: Facile hydrothermal construction of Nb_2CT_x/Nb_2O_5 as a hybrid anode material for high-performance Li-ion batteries. Chinese Chem. Lett. **31**(4), 1030–1033 (2020). https://doi.org/10.1016/j.cclet.2020.03.006
17. Wermuth, T.B., et al.: Microwave-synthesized $KNbO_3$ perovskites: photocatalytic pathway on the degradation of rhodamine B. Ceram. Int. **45**(May), 24137–24145 (2019). https://doi.org/10.1016/j.ceramint.2019.08.122
18. Ruwer, T.L., Venturini, J., Khan, S., Bergmann, C.P.: Quick synthesis of homogeneous Nb_2O_5 nanorod arrays via a microwave-assisted hydrothermal method. Mater. Lett. **265**, 127429 (2020). https://doi.org/10.1016/j.matlet.2020.127429
19. Braga, V.S., Garcia, F.A.C., Dias, J.A., Dias, S.C.L.: Copper oxide and niobium pentoxide supported on silica-alumina: Synthesis, characterization, and application on diesel soot oxidation. J. Catal. **247**(1), 68–77 (2007). https://doi.org/10.1016/j.jcat.2006.12.022

20. Liu, F., Xue, D.: Fabrication of Nb_2O_5 nanotrees with controlled branching degrees. Phys. Scr. T **T139** (2010). https://doi.org/10.1088/0031-8949/2010/T139/014074
21. Luo, H., Wei, M., Wei, K.: Synthesis of Nb_2O_5 nanosheets and its electrochemical measurements. Mater. Chem. Phys. **120**(1), 6–9 (2010). https://doi.org/10.1016/j.matchemphys.2009.10.053
22. Wen, H., Liu, Z., Wang, J., Yang, Q., Li, Y., Yu, J.: Facile synthesis of Nb_2O_5 nanorod array films and their electrochemical properties. Appl. Surf. Sci. **257**(23), 10084–10088 (2011). https://doi.org/10.1016/j.apsusc.2011.07.001
23. Li, C.C., Dou, J., Chen, L., Lin, J., Zeng, H.C.: Antisolvent precipitation for the synthesis of monodisperse mesoporous niobium oxide spheres as highly effective solid acid catalysts. Chem. Cat. Chem. **4**(10), 1675–1682 (2012). https://doi.org/10.1002/cctc.201100457
24. S.B.: Max-Planck-Institutfür Metallforschung, Institut für Sondermetalle, "Thermodynamische beschreibung der festen lösung von kohlenstoff, stickstoff und sauerstoff in niob und tantal. J. Less-Common Metals **14** (1967)
25. Galstyan, V., Comini, E., Faglia, G., Sberveglieri, G.: Synthesis of self-ordered and well-aligned Nb_2O_5 nanotubes. Cryst. Eng. Commun. **16**, 10273–10279 (2014). https://doi.org/10.1039/C4CE01540A
26. Roy, P., Berger, S., Schmuki, P.: TiO_2 Nanotubes: Synthesis and Applications (2011). http://dx.doi.org/https://doi.org/10.1002/anie.201001374

Nanomaterials for Magnetic Hyperthermia

Mariana Borges Polla, Oscar Rubem Klegues Montedo, and Sabrina Arcaro

Abstract Cancer is a disease with a tremendous mortality rate, in which the lack of early diagnosis, the cytotoxicity created by conventional techniques, and resistance to conventional therapies make its treatment difficult. In this sense, materials development on a nanometric scale has offered significant industrial and scientific advances, especially in biomedicine. In this chapter, we present the necessary characteristics and properties for nanomaterials in the field of magnetic hyperthermia. The basic concepts of magnetism suitable for the application and the structural and morphological characteristics necessary to apply nanomaterials in this field are addressed. Superparamagnetism is the main property to be achieved for applications in this area. Superparamagnetic nanoparticles can be oriented and located on the therapeutic target and heat up with the application of a magnetic field. From this understanding, it is concluded that magnetic hyperthermia is an up-and-coming technology for applications in cancer treatment, overcoming the limitations of conventional approaches.

Keywords Magnetic hyperthermia · Magnetic nanoparticles · Superparamagnetism

1 Introduction

Cancer is becoming the leading global health problem. Estimates by 2035, according to the World Health Organization (WHO), are that 24 million new cases and 14.5 million cancer-related deaths per year may appear in the world. Furthermore, cancer incidence and mortality have been increasing due to aging, population growth, and the change in the distribution and prevalence of risk factors, especially those associated with socioeconomic development [1]. The term cancer is designated to a set of more than 100 diseases that have in common the disordered growth of cells, which spread rapidly to tissues and organs. By adopting a new resolution on cancer,

M. B. Polla (✉) · O. R. K. Montedo · S. Arcaro
Grupo de Biomateriais E Materiais Nanoestruturados, Programa de Pós-Graduação Em Ciência e Engenharia de Materiais, Universidade Do Extremo Sul Catarinense, Av. Universitária 1105, P.O. Box 3167, Criciúma 88806-000, Brazil

A. Kopp Alves (ed.), *Technological Applications of Nanomaterials*,
Engineering Materials, https://doi.org/10.1007/978-3-030-86901-4_10

health leaders from several countries—including Brazil—reaffirmed the importance of disease control as a critical priority for health and development at the 70th World Health Assembly in Geneva.

Furthermore, the current United Nations (UN) Global Plan of Action on chronic non-communicable diseases and the ambitious goals of the United Nations Sustainable Development Goals for 2030, including goal 3 (good health and well-being), create a scenario of opportunities to invest in fighting cancer, one of the main on non-communicable chronic diseases. Thus, goal 3.4 of goal 3 mentions that: "By 2030, reduce premature mortality from non-communicable diseases by one third through prevention and treatment, and promote mental health and well-being". There are several types of cancer treatments, and some are shown in Fig. 1. Conventional treatments involve surgery, radiotherapy, chemotherapy, or a combination of these [2]. These treatments are poorly selective and end up causing toxic effects to surrounding healthy tissues, as well as numerous side effects [2, 3]. However, none of them alone can completely eradicate the malignant tumor tissue. Surgical treatment offers the best chance of cure. Therefore, there are several efforts to develop new treatments for these diseases, with fewer side effects, high safety, and better efficiency. However, few novelties were added to conventional therapy.

Hyperthermia is a promising treatment proposal, where tumor cells are affected by local temperature elevation in the order of 43 to 45 °C. Under these conditions, cancer cells are more sensitive to the cytotoxic effect of heat, and therefore, hyperthermia causes cell death of tumor cells with no effect on healthy cells [3, 4]. The temperature increase required by hyperthermia can be achieved, among other methods, by the use of magnetic nanoparticles. From the control of size and shape, high saturation magnetization, and adjustment of functionalization, it is possible to produce potential magnetic nanoparticles for application in cancer treatment. These materials have

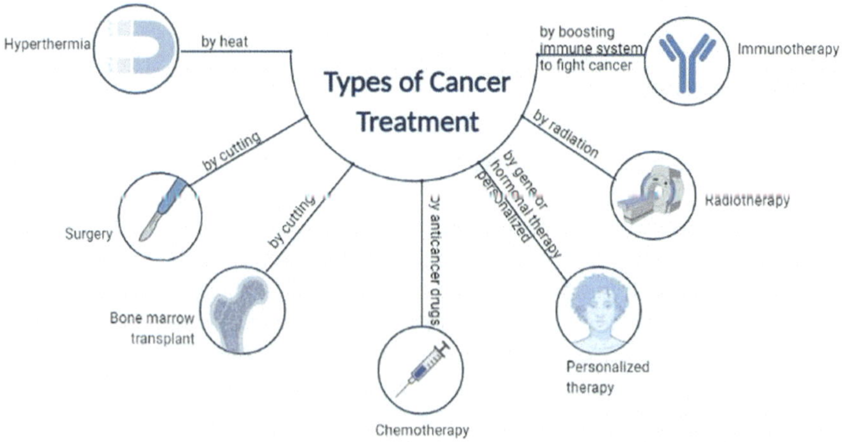

Fig. 1 Several types of cancer treatments

new characteristics that enable a wide variety of innovative applications and provide desirable characteristics of biodegradability and biocompatibility.

2 Hipertermia Magnética

Hyperthermia word derives from the Greek "hyper" (which means up or above) and "therme" (heat). Hyperthermia is a treatment in which a region of the body affected by a tumor is exposed to high temperatures to destroy cancer cells or make them more sensitive to the effects of radiation and anti-cancer drugs. Cancer cells are more sensitive to heat than healthy ones. These cells die at temperatures above 42 °C because the natural enzymatic process that maintains the living cells is disrupted [5].

Different approaches have been used to generate hyperthermia in tumor regions, such as techniques involving laser, microwave, radiofrequency, and ultrasound [6]. However, as the energy source in these techniques is external, most of the energy is dissipated in healthy cells along the radiation path. This lack of selectivity causes damage to healthy tissue [4].

In this context, superparamagnetic nanoparticles have been researched as agents of hyperthermia to offer a localized treatment. When subjected to an alternating magnetic field, these nanoparticles release heat due to Néel or Brown relaxation mechanisms [4]. Local heating can reach temperatures in the order of 43 to 45 °C, depending on the intensity and frequency of oscillation of the magnetic field and the intrinsic characteristics of the particles [7, 8].

The treatment of magnetic hyperthermia involves injecting biocompatible magnetic nanoparticles directly into the tumor or into the blood vessels that supply it. After reaching the therapeutic target, an alternating magnetic field is applied so that the magnetic nanoparticles generate heat through superparamagnetic behavior. In this sense, during the alternating magnetic field application, the particles' magnetization direction is continuously reversed, generating the heat needed to lyse the tumor cells. [9, 10]. If this temperature can be maintained in the preferential range of 42–44 °C for 30 min or more, the tumor may be totally or partially destroyed. Minimizing damage to surrounding normal tissues makes magnetohyperthermia a promising technique for the treatment of various cancers. [11]. The nanoparticles must be dispersed in a solution with physiological pH (ferrofluids) and are stable for this application. To induce target-specific hyperthermia, magnetic nanoparticles are guided and retained in the tumor region by an external magnetic field gradient after administration.

The advantages of using magnetic oxide nanoparticles in biomedical applications involve [12–14]:

1 The ability to interact with biological entities of interest due to size control;
2 The ability to respond to an external magnetic field gradient. As human tissues are penetrable to the magnetic field, nanoparticles can be transported to tissues of interest, such as tumors, and present a specific action;

3 A large surface area/volume ratio, which allows for surface modification. As the surface area increases, the number of atoms on the surface also increases, resulting in high chemical reactivity and more significant surface adsorption. As a result, the surface can be modified to improve stability, prevent particle aggregation, improve biocompatibility and dispersion, thus increasing circulation time in vivo and, at the same time, ameliorating its cytotoxicity.

So, for biomedical applications such as hyperthermia, magnetic nanoparticles must have a high magnetization value, superparamagnetic behavior at room temperature, shape control, and a narrow range of size distribution [14].

3 Appropriate Magnetism for the Application of Hyperthermia

Understanding fundamental concepts about magnetic particles and the effect of physical, chemical, and thermal characteristics on their properties is crucial to designing these materials with properties suitable for hyperthermic cancer treatment applications. All materials interact with magnetic fields and can be classified into several categories depending on their magnetic behaviour. They are classified based on the arrangement of magnetic dipoles in the absence or presence of an external magnetic field. Thus, they can be diamagnetic, paramagnetic, ferromagnetic, ferrimagnetic, and antiferromagnetic. The ways of aligning the magnetic moments within a crystal are shown in the simplified representation of Fig. 2, where H represents the applied magnetic field.

Diamagnetics do not show magnetic dipoles in the absence of a magnetic field but generate magnetic dipoles in the opposite direction to the applied field. This is because the insertion of electrons, which surround the nucleus of an atom in a magnetic field, promotes the action of the Lorentz force and reaches an induced field in the opposite direction to the original. This interaction causes the electrons to be repelled by the applied field, causing diamagnetism [15]. Due to the negligible relative intensity of this effect, exclusively materials that do not present another magnetic interaction are called diamagnetic.

Materials that are characterized by incomplete atomic or molecular orbitals are related differently. So, in paramagnetic materials, randomly oriented dipoles in the absence of a magnetic field are aligned in the same direction by interacting with an external magnetic field. According to the Pauli Exclusion Principle, paired electrons that exhibit reverse spins cancel each other out. However, when unpaired, electrons are free to align their spin in all directions. Thus, when introduced into a magnetic field, these electrons line up parallel to the applied field. Materials present paramagnetism with incomplete atomic orbitals—e.g., several coordination complexes—and those with incomplete molecular orbitals, such as the molecule of O_2.

Ferromagnetic materials have permanent magnetic dipoles, both in the absence and presence of an external magnetic field. In this case, the spins are already naturally

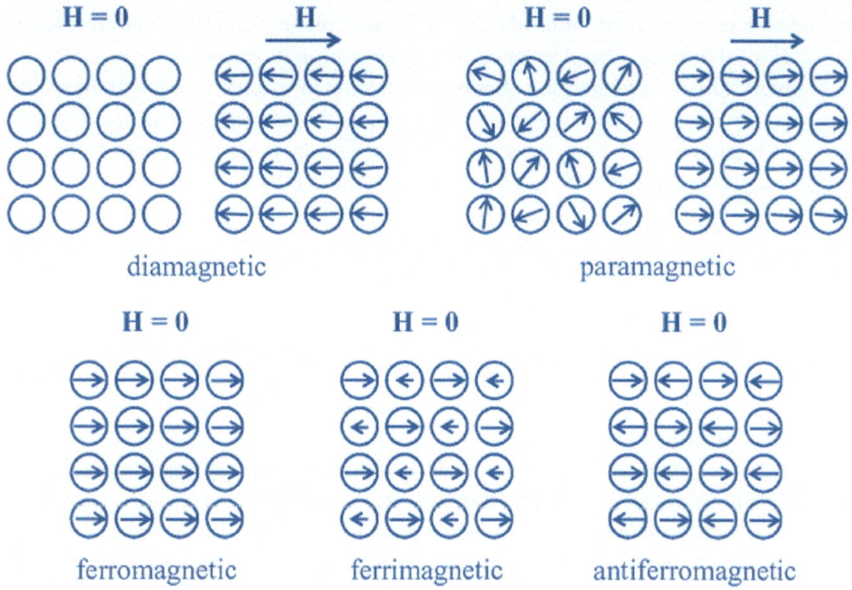

Fig. 2 Representation of the arrangement of magnetic dipoles of different materials in the absence and presence of an applied magnetic field (H)

aligned through spontaneous magnetization. As long as all spins in the crystallite are aligned in the same direction, the magnetic moments are added up, and a significant macroscopic effect can be observed. Among the few materials that present ferromagnetism, iron, cobalt, and rare earth can be mentioned [16]. Ferromagnetism exists exclusively below a Curie limit temperature, above which the spontaneous alignment of magnetic moments disappears due to thermal energy input.

In ferrimagnetic materials, some weak magnetic dipoles also exist in antiparallel to neighbouring strong dipoles without a magnetic field. Neighbouring dipoles are antiparallel and mutually neutralize in the absence of a magnetic field in the antiferromagnetic materials. At moments with different intensities (e.g., two distinct magnetic cations in the same unit cell), the net magnetization is different from zero. Materials that exhibit this behaviour are called ferrimagnetic. Fe_3O_4 is an example of a ferrimagnetic structure. The sublattices of Fe^{3+} and Fe^{2+} ions are arranged in an antiparallel way; however, because their magnetic moments are different, they do not cancel each other, and an intense magnetization resulting from this process can be visualized [15].

Based on the field required for demagnetization, materials that show spontaneous magnetization fall into two other categories: hard magnets and soft magnets. This factor, called coercivity (HC), is characterized by the field strength required to nullify the magnetism of the material after it has been brought to magnetic saturation (Ms). Materials with a high coercivity (i.e., "permanent" magnetization) are called magnetically hard and apply, for example, in data recording. On the other hand, materials that lose their magnetization easily are called soft magnets and are mainly used in applications that need a brief reversal of magnetization direction.

Magnetic anisotropy energy makes magnetization in specific alignment directions of magnetic moments more energetically favourable than others. Magnetic anisotropy can be estimated from the magnetic properties, which are determined in the vibrating sample magnetometer (VSM) test, using Eqs. 1 and 2 [17]:

$$M = M_s\left(1 - b/H^2\right) \quad (1)$$

$$K_{eff} = \mu_0 M_s (15b/4)^{1/2} \quad (2)$$

where MS is the saturation magnetization, and b is correlated with the effect of crystalline anisotropy.

In the case of uniaxial magnetic crystals, knowing the fit parameter b, the effective anisotropic constant (K_{eff}) can be estimated using Eq. 2.

When iron or ferrimagnetic particles are small enough to be composed of only one magnetic domain, the phenomenon of superparamagnetism can occur. In the absence of domain walls and under a crystallite limit size, the electronic spins vary their alignment under the thermal influence [18]. However, if the measured time of magnetism is long enough, an example of most measurement techniques, these particles are defined by null net magnetization [19].

The most relevant phenomena of superparamagnetism need to be well understood. A macroscopic magnetic particle is composed of thousands of magnetic domains separated by boundaries called domain walls, forming crystallites (Fig. 3). The domain wall can be considered a type of defect in the material. It needs the energy to form and has significant width (up to hundreds of nanometers). This wall is formed

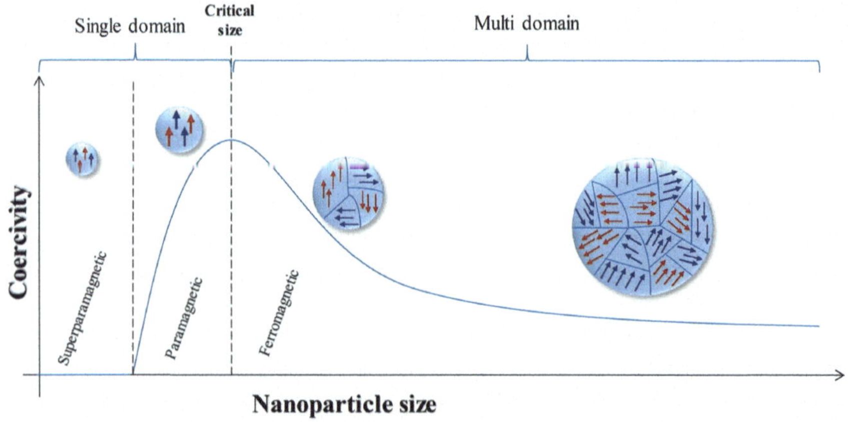

Fig. 3 Correlation between particle size, coercive field, and magnetic domain structures. Reprinted from [20] with permission Elsevier

from the balance between magnetostatic energy and domain wall energy. Magnetostatic energy is dependent on the volume of materials. The domain wall energy increases proportionally with the interface area between the domains. In macroscopic particles then, the reversal of magnetization is controlled by the nucleation of movement of the domain walls in the material [20, 21].

When the particle size is reduced, the lowest energy state has a uniform magnetization, and the particle is composed of a single magnetic domain. More energy is needed below a critical size to form the domain walls than to support external magnetostatic energy in the single-domain state. Under the application of an external magnetic field, a particle in the single-domain state is uniformly magnetized, with all spins aligned in the direction of the applied magnetic field. In this case, the absence of domain walls allows the free rotation of the spins, responsible for the reversal of magnetization. This feature is crucial for the application of this type of material in magnetic hyperthermia [20].

The other essential phenomenon related to the size effect is the superparamagnetic threshold. As already mentioned, superparamagnetism can be explained by considering single domain particles isolated from each other. In this case, the energy needed to keep the magnetic moments aligned in a specific direction is dictated by the anisotropic magnetic energy of the particles. This relationship is expressed by:

$$E_\theta = K_{eff} V \sin^2\theta \qquad (3)$$

where K_{eff} is the anisotropy constant, V is the particle volume, and θ is the angle between the magnetization and the axis.

Thus, the energy barrier, defined by $K_{eff}V$, separates two energetically equivalent preferred directions of magnetization.

Then, when the particle volume is reduced, the thermal energy exceeds the energy barrier, and magnetization is easily reversed. In this case, the system behaves like a paramagnet rather than individual atomic magnetic moments, becoming superparamagnetic [20]. In summary, superparamagnetic nanoparticles become magnetic when an external magnetic field is applied. However, they can easily be reverted to non-magnetic when the magnetic field is removed. Therefore, superparamagnetic nanoparticles have zero coercivity and do not show hysteresis.

At room temperature, bulk magnetite shows ferrimagnetic behaviour, high Curie temperature (~850 K), high saturation magnetization (92 emu/g, T = 20 °C), and coercive field between 200 and 400 Oe. On the other hand, magnetite nanoparticles with a size below 20 nm exhibit superparamagnetic behaviour [22–25]. In this case, the saturation magnetization (Ms), remanent magnetization (Mr), and coercive field (Hc) values vary according to the size of the nanoparticles. Therefore, nanoparticles below 7 nm exhibit a low Ms value. On the other hand, magnetic nanoparticles with a size in the range of 14–16 nm have a higher Ms value, generating adequate heat for an efficient hyperthermia cancer treatment without affecting significant collaterals. The shape of the nanoparticles is also essential for efficient heat generation. Magnetic nanoparticles containing faces with a lower energetic surface and have lower K and

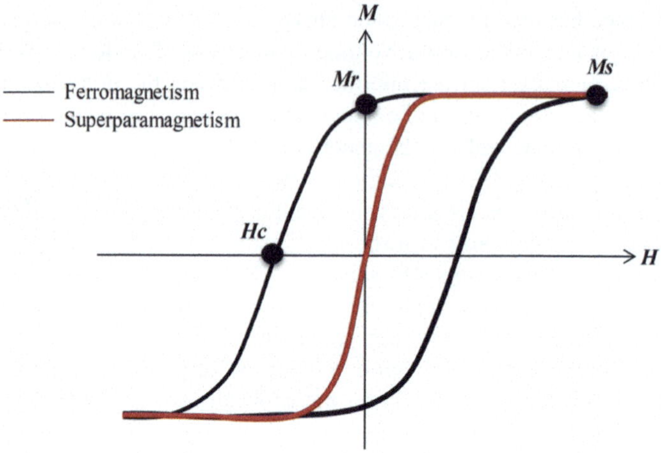

Fig. 4 Schematic representation of the hysteresis cycle for ferromagnetic and superparamagnetic materials. Reprinted from [20] with permission Elsevier

higher Ms. Therefore, these nanoparticles have shown greater heating efficiency. For example, spherical iron oxide nanoparticles had lower Ms values compared to the same cubic nanoparticles with an equivalent volume [12].

Typical hysteresis curves for magnetic materials are shown in Fig. 4. The magnetization curve as a function of the applied field presents zero remanent magnetization and coercive field in superparamagnetic materials. That is, the particle does not present magnetic hysteresis. This characteristic provides excellent advantages for applications in hyperthermia, as the particles lose magnetization after the end of the magnetic field application, eliminating the leading cause of agglomeration. Furthermore, it is possible to easily control the magnetic properties as they respond to an external magnetic field application [12].

4 Heat Generation Mechanisms of Magnetic Nanoparticles

In magnetic hyperthermia, the two main mechanisms responsible for generating heat from magnetic nanoparticles are hysteresis losses and magnetic relaxation losses (Néel or Brown relaxation). In nanoparticles that have a magnetic multidomain, the heating effect occurs due to hysteresis losses. Magnetic relaxation occurs mainly in single-domain or superparamagnetic nanoparticles, as illustrated in Fig. 5 [12].

Neel's relaxation is related to the change of direction of the magnetic moment. This relaxation mechanism is the only one that occurs when nanoparticles are immobilized in tumor cells [4]. When the magnetic field is removed, the nanoparticles' magnetic moments cease after a relaxation time (τ) due to thermal agitation (kT). In superparamagnetic nanoparticles, the energy barrier for magnetic relaxation is

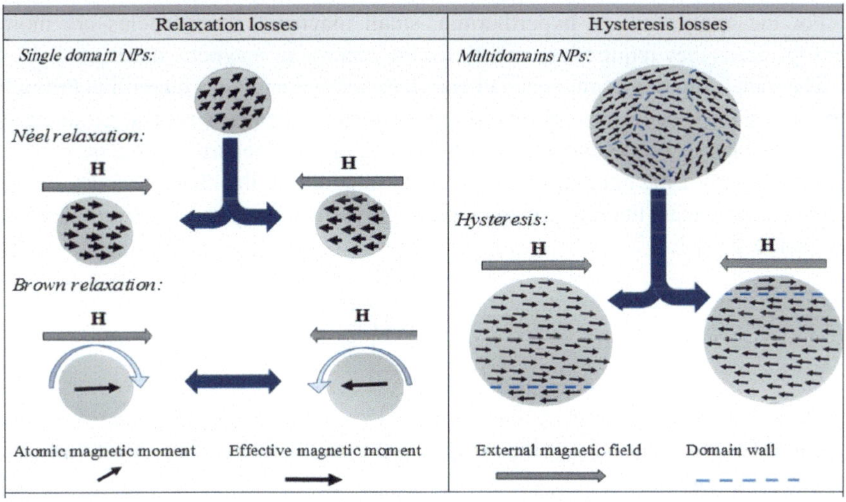

Fig. 5 Principais mecanismos de geração de calor via nanopartículas magnéticas sob campo magnético aplicado. Reprinted from [4] with permission Elsevier

described by KV, where K is the magnetic anisotropy and V is the particle volume. This relaxation time is called the Neel relaxation time (τN) and rapidly decreases with the volume of particles [4]. Néel's relaxation time can be calculated from Eq. 4.

$$\tau_N = \tau_0 e^{\frac{KV_M}{kT}} \tag{4}$$

where τ_0 is the characteristic frequency of inversion (~10–9 s), K is the magnetic anisotropy constant, VM is the particle volume, k is the Boltzmann constant, and T is the temperature.

In turn, the release of heat by Brown relaxation is related to the rotation of particles and collisions between them in a liquid medium. This mechanism occurs in magnetic fluid and is related to the viscosity of the carrier liquid (η). Brown's relaxation time can be calculated by Eq. 5.

$$\tau_B = \frac{3\eta V_h}{\kappa T}. \tag{5}$$

where Vh is the hydrodynamic volume of the particle and η is the viscosity of the magnetic fluid.

During magnetization reversal, particles take the easiest and shortest relaxation time path. As a result, the adequate relaxation time (τ) is given by Eq. 6.

$$\tau = \frac{\tau_B \tau_N}{\tau_B + \tau_N} \tag{6}$$

For the application of hyperthermia, small magnetic nanoparticles are more appropriate as they require less energy for the rotation of magnetic moments.

The transformation of magnetic energy into thermal energy is quantified through the value of the specific absorption rate (SAR), which depends on several parameters, including the size, shape, and magnetic properties of the nanoparticles, the frequency, and amplitude of the applied magnetic field. In addition, the characteristics of the coating material and the functionalization of the surface [12]. The SAR value can be calculated using Eq. 7.

$$SAR = c\left(\frac{dT}{dt}\right)\left(\frac{M}{m}\right) \tag{7}$$

where c is the specific heat of the sample, dT/dt is the heat release rate from the system, M is the total mass of the sample, and m is the mass of magnetic particles present in the sample.

To assess the intrinsic capacity of magnetic nanoparticles to generate heat, regardless of the intensity and frequency of the magnetic field. The value of the intrinsic power to loss (IPL) of the ferrofluid can be calculated using Eq. 8 [17].

$$IPL\left(nHm^2kg^{-1}\right) = \frac{SAR}{fxH^2} \tag{8}$$

where f is the frequency and H is the strength of the applied magnetic field.

It is worth noting that nanoparticles with a high SAR value are desirable for clinical applications, as they allow the use of a lower dosage of nanoparticles [12].

5 Nanopartículas Magnéticas Para Aplicação Em Hipertermia Magnética

Magnetic nanoparticles are of great interest in biomedicine due to their wide variety of applications. In recent years, one of the most challenging goals has been developing new strategies to adjust the unique properties of magnetic nanoparticles to improve their effectiveness in this field. One of the most promising applications in biomedicine is, in fact, the treatment of cancer by magnetic hyperthermia [12, 26].

The first advantage of using magnetic oxide nanoparticles for biomedical applications lies in the fact that it is possible to control the size of these structures by manipulating the synthesis parameters, from a few nanometers to hundreds of nanometers, which places them in smaller dimensions than cells (10–100 μm) or sizes comparable to viruses (20–450 nm), proteins (5–50 nm) or genes (2 nm wide and 10–100 nm long). This means that nanoparticles can interact with the biological entity of interest [13, 14]. The second advantage is that these particles are superparamagnetic. That is, they can respond to an external magnetic field gradient only when the field is applied.

As human tissues have penetrability to the magnetic field, these nanoparticles can be transported to tissues of interest and specific. Furthermore, superparamagnetic particles do not retain any remaining magnetization after removing the magnetic field, thus being of interest for in vivo applications [27]. This fact is fundamental, as paramagnetic particles (with a size greater than 20 nm) aggregate after exposure to a magnetic field and thus can cause problems such as embolisms. To optimize the balance between the contributions of Néel and Brown [28], the most suitable particle sizes for this application are around 15 nm.

As previously discussed, currently, iron oxides, magnetite (Fe_3O_4) [29, 30], and maghemite (-Fe_2O_3) [31, 32] are the only class of magnetic nanoparticles approved for in vivo applications by the FDA (Food and Drug Administration). These nanoparticles have the advantages of biocompatibility [33], low toxicity [34], and superparamagnetic behavior. In addition to the mentioned iron oxides, other types of ferrites are also being studied for application in hyperthermia, such as: $CoFe_2O_4$ [35–38], $NiFe_2O_4$ [39, 40], $CuFe_2O_4$ [41, 42], $ZnFe_2O_4$ [43, 44] e $MnFe_2O_4$ [45–47] and other oxides [48–50].

The material class of ferrites is materials based on mixed compounds of iron (III) oxide (Fe_2O_3) and one or more transition metal oxides. Its general formula is MFe_2O_4, where M is a divalent ion (Fe^{2+}, Co^{2+}, Ni^{2+}, Cu^{2+}, Zn^{2+} e Mn^{2+}). This structure is characterized by packing oxygen ions into a face-centered cubic arrangement. Among the oxygen ions are Fe^{3+} and M^{2+} ions occupying tetrahedral and octahedral interstices. The most general representation of this structure is presented by Eq. 9:

$$\left(M^{2+}_{(1-x)}Fe^{3+}_{x}\right)_A\left[Fe^{3+}_{(2-x)}M^{2+}_{x}\right]_B O^{2-}_{4} \tag{9}$$

where x is the metal ion occupancy parameter; parentheses are tetrahedral sites, and square brackets are octahedral sites.

When the M^{2+} ions occupy the tetrahedral sites and the Fe^{3+} the octahedral sites, we have a direct spinel-type crystal structure ($x = 0$). If the Fe^{3+} ions are evenly split between the tetrahedral and octahedral sites, and the divalent metal ions occupy the octahedral sites, the structure is called the inverse spinel ($x = 1$). In case these cations are distributed, we have a mixed spinel. Figure 6 is a representation of the magnetite unit cell (Fe_3O_4).

For practical application in vivo, the main obstacle is making the nanoparticles reach a particular location in the body, being stable in body fluids. As mentioned previously, one of the benefits of using magnetic nanoparticles is the use of localized magnetic field gradients to attract the particles to a particular selected location, where they remain for the duration of therapy and then can be removed [27]. The surface of these particles can be modified by creating some atomic layers composed of organic polymers (such as polylactic acid, dextran, PEG, PVP, PVA, and dendrimers) [52–56], metal (Au, Gd e Pt) [57–59], metal oxides (SiO_2, Al_2O_3, TiO_2) [60–62], fatty acids (oleic acid) or amino acids [63, 64]. Bioactive molecules such as biotin, avidin, carboxyl groups, carbodiimide, antibodies, hormones, folic acid, among others, can be treated on the surface of the nanoparticles to direct the nanoparticles

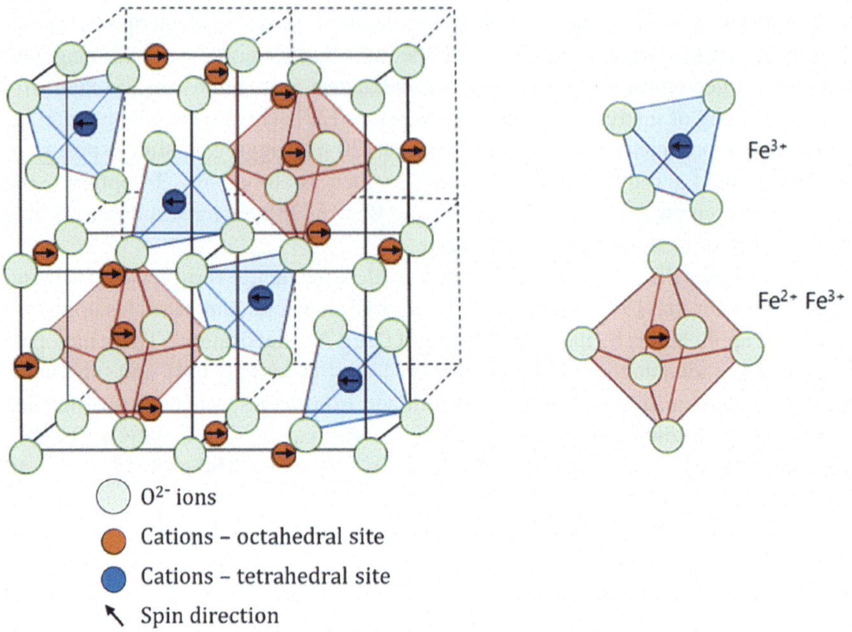

Fig. 6 Schematic representation of the crystal structure of magnetite, showing the position of the O^{2+} ions, the Fe^{2+} and Fe^{3+} cations, the octahedral and tetrahedral sites, and the direction of the spins. Reprinted from [51] with permission Elsevier

to the target cells [65, 66]. In addition to the surface modification promoting a more specific interaction of the particle with the tissue, cell, or molecule, surface coverage also increases the magnetic fluid's colloidal stability [27].

The different coatings provide the magnetic nanoparticles a significant improvement in biocompatibility and stabilize a liquid medium. However, after recoating, there is a decrease in magnetic properties, more specifically in Ms. Consequently, the SAR decreases. Therefore, it remains a challenge to produce functionalized magnetic nanoparticles that have a high SAR value.

6 Production of Magnetic Nanoparticles

Several synthesis methods have been developed to produce magnetic nanoparticles with good shape, composition, and size distribution control (Table 1). These nanoparticles can be produced using physical, chemical, or biological processes. The physical processes, known as top-down, reduce larger size materials to micro and nanoscale, starting from macroscopic materials. However, this process is time-consuming, involves high-cost technologies, and nanoparticles have a non-uniform size distribution.

Table 1 Synthesis conditions, particle size and nanoparticles saturation magnetization

Method	Synthesis conditions	Solvent	Size distribution	Size (nm)	Shape control	Magnetization (emu/g)	References
Coprecipitation	20–90 °C, 3–5 h; extreme pH conditions	Water	Large	15	Poor	50–60	[78–82]
Thermal decomposition	100–320 °C, inert atmosphere	Organic; diphenyl ether	Very close	5–15	Very good	70–90	[29, 74, 82, 83]
Microemulsion	20–80 °C, ambient conditions	Organic	Relatively close	<10	Very good		[77, 82, 84]
Hydrothermal	100–220 °C, hours-weeks, high pressures	Water / ethanol	Very close	20–60	Very good	20–80	[28, 75, 82, 85, 86]
Sol–gel	40–120 °C; 1- 4 h; Thermal treatment 300 °C	Water; ethylene glycol	Close	8,5–15,5	Very good	30–60	[48, 51, 82, 87–91]
Sonochemical	Room Temperature; 3–30 min ultrasound high-frequency	Water	Wide	10–50	Good	10–60	[82, 83, 92, 93]
Electrochemical deposition	high acidity	Phosphoric acid and sulfuric acid	Close	8–10	Regular	70	[82, 83]
Polyol Method	100–300 °C	Octyl ether	Very close	5	Very good		[82, 94, 95]

Chemical and biological processes are bottom-up processes and consist of methods that build the nanostructure atom by atom. The magnetic properties of nanoparticles are strongly affected by the synthesis process. Therefore, chemical routes are used when it is desired to obtain materials with controlled size, shape, and chemical composition. Among the chemical methods used are sol–gel synthesis [67–70], coprecipitation [71–73], thermal decomposition [74], hydrothermal or solvothermal [41, 75], microemulsion [76, 77], and others.

7 Conclusions

This chapter focuses on the characteristics and properties needed for nanomaterials in the field of magnetic hyperthermia. From the proper selection of the synthesis method, size and shape control, saturation magnetization, and functionalization adjustment, it is possible to produce potential magnetic nanoparticles for application in cancer treatment. Furthermore, these materials have new characteristics that enable a wide variety of innovative applications and provide desirable characteristics of biodegradability and biocompatibility.

Although the clinical application of magnetic devices may not be that close, a remarkable development is expected in the coming years. Successful advances in this area will increase the growth of cancer therapy in the biomedical industry and may finally enable the use of this technology in cancer treatment.

Acknowledgements The authors are very grateful to Coordination for the Improvement of Higher Education Personnel (Coordenação de Aperfeiçoamento de Pessoal de Nível Superior, CAPES/Brazil), National Council of Technological and Scientific Development (Conselho Nacional de Desenvolvimento Científico e Tecnológico, CNPq/Brazil; process n. 161197/2020-5; 307761/2019-3) and Fundação de Amparo à Pesquisa e Inovação do Estado de Santa Catarina (Fapesc/Brazil) for supporting this work.

References

1. National Cancer Institute: Why Global Cancer Research Is Critical to Progress against the Disease (2021)
2. INCA: O que é câncer? In: 03/04/2019 (2019)
3. Yu, X., Ding, S., Yang, R., et al.: Research progress on magnetic nanoparticles for magnetic induction hyperthermia of malignant tumor. Ceram. Int. (2020). https://doi.org/10.1016/j.ceramint.2020.11.049
4. Beik, J., Abed, Z., Ghoreishi, F.S., et al.: Nanotechnology in hyperthermia cancer therapy: From fundamental principles to advanced applications. J. Control Release **235**, 205–221 (2016). https://doi.org/10.1016/j.jconrel.2016.05.062
5. Rajan, S.A., Sahu, N.K.: Inductive calorimetric assessment of iron oxide nano-octahedrons for magnetic fluid hyperthermia. Colloids Surf. Physicochem. Eng. Asp **603**, 125210 (2020). https://doi.org/10.1016/j.colsurfa.2020.125210

6. Danewalia, S.S., Singh, K.: Bioactive glasses and glass–ceramics for hyperthermia treatment of cancer: state-of-art, challenges, and future perspectives. Mater. Today Bio. **10**, 100100 (2021). https://doi.org/10.1016/j.mtbio.2021.100100
7. Bañobre-López, M., Teijeiro, A., Rivas, J.: Magnetic nanoparticle-based hyperthermia for cancer treatment. Reports Pract. Oncol. Radiother. (2013)
8. Hedayatnasab, Z, Abnisa, F., Mohd Ashri Wan Daud, W.: Review on magnetic nanoparticles for magnetic nanofluid hyperthermia application (2017). https://doi.org/10.1016/j.matdes.2017.03.036
9. Lima-Tenório, M.K.,, Gómez Pineda, E.A., Ahmad, N.M. et al.: Magnetic nanoparticles: In Vivo Cancer Diagnosis And Therapy. Int. J. Pharm. (2015)
10. Dey, C., Ghosh, A., Ahir, M., et al.: Improvement of anticancer drug release by cobalt ferrite magnetic nanoparticles through combined ph and temperature responsive technique. Chem. Phys. Chem. **19**, 2872–2878 (2018). https://doi.org/10.1002/cphc.201800535
11. Chatterjee, D.K., Diagaradjane, P., Krishnan, S.: Nanoparticle-mediated hyperthermia in cancer therapy. Ther. Deliv. **2**, 1001–1014 (2011). https://doi.org/10.4155/tde.11.72
12. Das, P., Colombo, M., Prosperi, D.: Recent advances in magnetic fluid hyperthermia for cancer therapy. Colloids Surf. B Biointerfaces **174**, 42–55 (2019). https://doi.org/10.1016/J.COLSURFB.2018.10.051
13. Sargentelli, V., Ferreira, A.P.: Nanopartícilas Magnéticas: O Cobalto. Eclet Quim **35**,153–163. https://doi.org/10.1590/S0100-46702010000400020
14. Fontanive, V.C.P., Khalil, N.M., Cotica, L.F., Mainardes, R.M.: Aspectos físicos e biológicos de nanopartículas de ferritas magnéticas. Rev Ciencias Farm Basica e Apl **35**, 549–558 (2014)
15. Ibach H, Lüth H (2010) Solid-state physics: An introduction to principles of materials science
16. Magnetism in the solid state. Mater Today (2013). https://doi.org/10.1016/s1369-7021(03)00433-4
17. Fotukian, S.M., Barati, A., Soleymani, M., Alizadeh, A.M.: Solvothermal synthesis of $CuFe_2O_4$ and Fe3O4 nanoparticles with high heating efficiency for magnetic hyperthermia application. J. Alloys Compd. 152548 (2019). https://doi.org/10.1016/j.jallcom.2019.152548
18. Néel, L., Louis Néel Professeur à la, P.M.: Théorie du traînage magnétique des substances massives dans le domaine de Rayleigh Théorie Du Traînage Magnétique Des Substances Massives Dans Le Domaine De Rayleigh. J. Phys. Radium. **11** (1950). https://doi.org/10.1051/jphysrad:01950001102049001
19. Mohn, P.: Magnetism in the Solid State : An Introduction. Springer (2006)
20. Soares, P.I.P., Romão, J., Matos, R., et al.: Design and engineering of magneto-responsive devices for cancer theranostics: Nano to macro perspective. Prog. Mater. Sci. **116**, 100742 (2021). https://doi.org/10.1016/j.pmatsci.2020.100742
21. Rahmani, R., Gharanfoli, M., Gholamin, M., et al.: Plant-mediated synthesis of superparamagnetic iron oxide nanoparticles (SPIONs) using aloe vera and flaxseed extracts and evaluation of their cellular toxicities. Ceram Int. **46**, 3051–3058 (2020). https://doi.org/10.1016/j.ceramint.2019.10.005
22. Sawatzky, G.A., Van Der Woude, F., Morrish, A.H.: Mössbauer study of several ferrimagnetic spinels. Phys. Rev. **187**, 747–757 (1969). https://doi.org/10.1103/PhysRev.187.747
23. Bushkova, V.S., Yaremiy, I.P.: Magnetic, electric, mechanical, and optical properties of NiCrxFe2−xO4ferrites. J. Magn. Magn. Mater. **461**, 37–47 (2018). https://doi.org/10.1016/j.jmmm.2018.04.025
24. Kefeni, K.K., Msagati, T.A.M., Mamba, B.B.: Ferrite nanoparticles: Synthesis, characterisation, and applications in electronic device. Mater. Sci. Eng. B Solid-State Mater. Adv. Technol. (2017)
25. Laurent, S., Dutz, S., Häfeli, U.O., Mahmoudi, M.: Magnetic fluid hyperthermia: Focus on superparamagnetic iron oxide nanoparticles. Adv. Colloid Interface Sci. (2011)
26. Dadfar, S.M., Roemhild, K.,, Drude, N.I. et al.: Iron oxide nanoparticles: Diagnostic, therapeutic and theranostic applications. Adv. Drug Deliv. Rev. (2019)
27. Berry, C.C., Curtis, A.S.G.: Functionalisation of magnetic nanoparticles for applications in biomedicine. J. Phys. D Appl. Phys. **36**, R198–R206 (2003). https://doi.org/10.1088/0022-3727/36/13/203

28. Lévy, M., Wilhelm, C., Siaugue, J.M., et al.: Magnetically induced hyperthermia: Size-dependent heating power of γ-Fe_2O_3 nanoparticles. J. Phys. Condens. Matter. (2008). https://doi.org/10.1088/0953-8984/20/20/204133
29. Gawali, S.L., Shelar, S.B., Gupta, J., et al.: Immobilization of protein on Fe3O4 nanoparticles for magnetic hyperthermia application. Int. J. Biol. Macromol. **166**, 851–860 (2021). https://doi.org/10.1016/j.ijbiomac.2020.10.241
30. Asgari, M., Miri, T., Soleymani, M., Barati, A.: A novel method for in situ encapsulation of curcumin in magnetite-silica core-shell nanocomposites: A multifunctional platform for controlled drug delivery and magnetic hyperthermia therapy. J. Mol. Liq. **324**, 114731 (2021). https://doi.org/10.1016/j.molliq.2020.114731
31. Ge, X., Fu, M., Kong, X.: Atomic layer deposition of γ-Fe_2O_3 nanoparticles on multi-wall carbon nanotubes for magnetic drug delivery and liver cancer treatment. Ceram. Int. **46**, 26557–26563 (2020). https://doi.org/10.1016/j.ceramint.2020.07.123
32. Shaw, S.K., Kailashiya, J., Gangwar, A., et al.: γ-Fe_2O_3 nanoflowers as efficient magnetic hyperthermia and photothermal agent. Appl. Surf. Sci. **560**, 150025 (2021). https://doi.org/10.1016/j.apsusc.2021.150025
33. Wang, W., Li, F., Li, S., et al.: M2 macrophage-targeted iron oxide nanoparticles for magnetic resonance image-guided magnetic hyperthermia therapy. J. Mater. Sci. Technol. **81**, 77–87 (2021). https://doi.org/10.1016/j.jmst.2020.11.058
34. Khmara, I., Strbak, O., Zavisova, V., et al.: Chitosan-stabilized iron oxide nanoparticles for magnetic resonance imaging. J. Magn. Magn. Mater. **474**, 319–325 (2019). https://doi.org/10.1016/j.jmmm.2018.11.026
35. Manohar, A., Geleta, D.D., Krishnamoorthi, C., Lee, J.: Synthesis, characterization and magnetic hyperthermia properties of nearly monodisperse $CoFe_2O_4$ nanoparticles. Ceram. Int. **46**, 28035–28041 (2020). https://doi.org/10.1016/j.ceramint.2020.07.298
36. Munjal, S., Khare, N., Sivakumar, B., Sakthikumar, D.N.: Citric acid coated $CoFe_2O_4$ nanoparticles transformed through rapid mechanochemical ligand exchange for efficient magnetic hyperthermia applications. J. Magn. Magn. Mater. **477**, 388–395 (2019). https://doi.org/10.1016/j.jmmm.2018.09.007
37. Phong, P.T., Phuc, N.X., Nam, P.H., et al.: Size-controlled heating ability of $CoFe_2O_4$ nanoparticles for hyperthermia applications. Phys. B Condens. Matter. **531**, 30–34 (2018). https://doi.org/10.1016/j.physb.2017.12.010
38. Suleman, M., Riaz, S.: In silico study of hyperthermia treatment of liver cancer using core-shell $CoFe_2O_4$@$MnFe_2O_4$ magnetic nanoparticles. J. Magn. Magn. Mater. **498**, 166143 (2020). https://doi.org/10.1016/j.jmmm.2019.166143
39. Egizbek, K., Kozlovskiy, A.L., Ludzik, K., et al.: Stability and cytotoxicity study of $NiFe_2O_4$ nanocomposites synthesized by coprecipitation and subsequent thermal annealing. Ceram. Int. **46**, 16548–16555 (2020). https://doi.org/10.1016/j.ceramint.2020.03.222
40. Lemine, O.M., Madkhali, N., Hjiri, M., et al.: Comparative heating efficiency of hematite (α-Fe_2O_3) and nickel ferrite nanoparticles for magnetic hyperthermia application. Ceram. Int. **46**, 28821–28827 (2020). https://doi.org/10.1016/j.ceramint.2020.08.047
41. Fotukian, S.M., Barati, A., Soleymani, M., Alizadeh, A.M.: Solvothermal synthesis of $CuFe_2O_4$ and Fe_3O_4 nanoparticles with high heating efficiency for magnetic hyperthermia application. J. Alloys Compd. **816**, 152548 (2020). https://doi.org/10.1016/j.jallcom.2019.152548
42. Kombaiah, K., Vijaya, J.J., Kennedy, L.J., et al.: Conventional and microwave combustion synthesis of optomagnetic $CuFe_2O_4$ nanoparticles for hyperthermia studies. J. Phys. Chem. Solids **115**, 162–171 (2018). https://doi.org/10.1016/j.jpcs.2017.12.024
43. Kerroum, M.A.A., Essyed, A., Iacovita, C., et al.: The effect of basic pH on the elaboration of $ZnFe_2O_4$ nanoparticles by coprecipitation method: Structural, magnetic and hyperthermia characterization. J. Magn. Magn. Mater. **478**, 239–246 (2019). https://doi.org/10.1016/j.jmmm.2019.01.081
44. Amiri, M., Gholami, T., Amiri, O., et al.: The magnetic inorganic-organic nanocomposite based on $ZnFe_2O_4$-Imatinib-liposome for biomedical applications, in vivo and in vitro study. J. Alloys Compd. **849**, 156604 (2020). https://doi.org/10.1016/j.jallcom.2020.156604

45. Patade, S.R., Andhare, D.D., Somvanshi, S.B., et al.: Self-heating evaluation of superparamagnetic $MnFe_2O_4$ nanoparticles for magnetic fluid hyperthermia application towards cancer treatment. Ceram. Int. **46**, 25576–25583 (2020). https://doi.org/10.1016/j.ceramint.2020.07.029
46. Mondal, D.K., Borgohain, C., Paul, N., Borah, J.P.: Improved heating efficiency of bifunctional $MnFe_2O_4$/ZnS nanocomposite for magnetic hyperthermia application. Phys. B Condens. Matter. **567**, 122–128 (2019). https://doi.org/10.1016/j.physb.2018.11.068
47. Mondal, D.K., Borgohain, C., Paul, N., Borah, J.P.: Tuning hyperthermia efficiency of $MnFe_2O_4$/ZnS nanocomposites by controlled ZnS concentration. J. Mater. Res. Technol. **8**, 5659–5670 (2019). https://doi.org/10.1016/j.jmrt.2019.09.034
48. Tonelli, A.M., Venturini, J., Arcaro, S., et al.: Novel core-shell nanocomposites based on TiO_2-covered magnetic Co_3O_4 for biomedical applications. J. Biomed. Mater. Res. - Part B Appl. Biomater. **108**, 1879–1887 (2020). https://doi.org/10.1002/jbm.b.34529
49. Avancini, T.G., Souza, M.T., de Oliveira, A.P.N., et al.: Magnetic properties of magnetite-based nano-glass-ceramics obtained from a Fe-rich scale and borosilicate glass wastes. Ceram. Int. **45**, 4360–4367 (2019). https://doi.org/10.1016/j.ceramint.2018.11.111
50. Phumying, S., Labuayai, S., Swatsitang, E., et al.: Nanocrystalline spinel ferrite (MFe2O4, M = Ni Co, Mn, Mg, Zn) powders prepared by a simple aloe vera plant-extracted solution hydrothermal route. Mater. Res. Bull. **48**, 2060–2065 (2013). https://doi.org/10.1016/J.MATERRESBULL.2013.02.042
51. Vitor, P.A.M., Venturini, J., da Cunha, J.B.M., Bergmann, C.P.: The influence of cation distribution on the magnetic properties of mixed $Co_{1-y}NiyFe_2O_4$ nanoferrites produced by the sol-gel method. J. Alloys Compd. (2021). https://doi.org/10.1016/j.jallcom.2020.156799
52. Shaterabadi, Z., Nabiyouni, G., Soleymani, M.: Optimal size for heating efficiency of superparamagnetic dextran-coated magnetite nanoparticles for application in magnetic fluid hyperthermia. Phys. C Supercond. Appl. **549**, 84–87 (2018). https://doi.org/10.1016/j.physc.2018.02.060
53. Asri, N.S., Tetuko, A.P., Esmawan, A., et al.: Syntheses of ferrofluids using polyethylene glycol (PEG) coated magnetite (Fe_3O_4), citric acid, and water as the working liquid in a cylindrical heat pipe. Nano-Struct. Nano-Objects **25**, 100654 (2021). https://doi.org/10.1016/j.nanoso.2020.100654
54. Jouyandeh, M., Ganjali, M.R., Ali, J.A., et al.: Curing epoxy with polyvinylpyrrolidone (PVP) surface-functionalized $Ni_xFe_{3-x}O_4$ magnetic nanoparticles. Prog. Org. Coat. **136**, 105259 (2019). https://doi.org/10.1016/j.porgcoat.2019.105259
55. Shagholani, H., Ghoreishi, S.M., Mousazadeh, M.: Improvement of interaction between PVA and chitosan via magnetite nanoparticles for drug delivery application. Int. J. Biol. Macromol. **78**, 130–136 (2015). https://doi.org/10.1016/j.ijbiomac.2015.02.042
56. Pan, B.-F., Gao, F., Ao, L.-M.: Investigation of interactions between dendrimer-coated magnetite nanoparticles and bovine serum albumin. J. Magn. Magn. Mater. **293**, 252–258 (2005). https://doi.org/10.1016/j.jmmm.2005.02.018
57. Han, Y., Lei, S., Lu, J., et al.: Potential use of SERS-assisted theranostic strategy based on Fe_3O_4/Au cluster/shell nanocomposites for bio-detection, MRI, and magnetic hyperthermia. Mater. Sci. Eng. C **64**, 199–207 (2016). https://doi.org/10.1016/j.msec.2016.03.090
58. Hirosawa, F., Iwasaki, T.: A comparative study of the magnetic induction heating properties of rare earth (RE = Y, La, Ce, Pr, Nd, Gd and Yb)-substituted magnesium–zinc ferrites. Solid State Sci. **118**, 106655 (2021). https://doi.org/10.1016/j.solidstatesciences.2021.106655
59. Koutsoumbou, X., Tsiaoussis, I., Bulai, G.A., et al.: $CoFe_{2-x}RE_xO_4$ (RE=Dy, Yb, Gd) magnetic nanoparticles for biomedical applications. Phys. B Condens. Matter. **606**, 412849 (2021). https://doi.org/10.1016/j.physb.2021.412849
60. Yu, X., Wang, L., Li, K. et al.: Tuning dipolar effects on magnetic hyperthermia of $Zn_{0.3}Fe_{2.7}O_4/SiO_2$ nanoparticles by silica shell. J. Magn. Magn. Mater. **521**, 167483 (2021). https://doi.org/10.1016/j.jmmm.2020.167483
61. Vasilakaki, M., Gemenetzi, F., Devlin, E., et al.: Size effects on the magnetic behavior of γ-Fe_2O_3 core/SiO_2 shell nanoparticle assemblies. J. Magn. Magn. Mater. **522**, 167570 (2021). https://doi.org/10.1016/j.jmmm.2020.167570

62. Islam, M.S., Kusumoto, Y., Abdulla-Al-Mamun, M., Horie, Y.: Photocatalytic and AC magnetic-field induced enhanced cytotoxicity of Fe_3O_4–TiO_2 core-shell nanocomposites against HeLa cells. Catal. Commun. **16**, 39–44 (2011). https://doi.org/10.1016/j.catcom.2011.08.039
63. Zhang, L., He, R., Gu, H.-C.: Oleic acid coating on the monodisperse magnetite nanoparticles. Appl. Surf. Sci.. **253**, 2611–2617 (2006). https://doi.org/10.1016/j.apsusc.2006.05.023
64. Urian, Y.A., Atoche-Medrano, J.J., Quispe, L.T., et al.: Study of the surface properties and particle-particle interactions in oleic acid-coated Fe3O4 nanoparticles. J. Magn. Magn. Mater. **525**, 167686 (2021). https://doi.org/10.1016/j.jmmm.2020.167686
65. Gholibegloo, E., Mortezazadeh, T., Salehian, F., et al.: Folic acid decorated magnetic nanosponge: An efficient nanosystem for targeted curcumin delivery and magnetic resonance imaging. J. Colloid. Interface Sci. **556**, 128–139 (2019). https://doi.org/10.1016/j.jcis.2019.08.046
66. Ivanova, A.V., Nikitin, A.A., Gabashvily, A.N., et al.: Synthesis and intensive analysis of antibody labeled single core magnetic nanoparticles for targeted delivery to the cell membrane. J. Magn. Magn. Mater. **521**, 167487 (2021). https://doi.org/10.1016/j.jmmm.2020.167487
67. Lemine, O.M., Omri, K., Iglesias, M., et al.: γ-Fe_2O_3 by sol–gel with large nanoparticles size for magnetic hyperthermia application. J. Alloys Compd. **607**, 125–131 (2014). https://doi.org/10.1016/j.jallcom.2014.04.002
68. PhD, S.O.A., Alnasir, M.H., Botha, S., et al.: The role of polyethylene glycol on the microstructural, magnetic, and specific absorption rate in thermoablation properties of Mn-Zn ferrite nanoparticles by sol–gel protocol. Eur. Polym. J. **132**, 109739 (2020). https://doi.org/10.1016/j.eurpolymj.2020.109739
69. Jasso-Terán, R.A., Cortés-Hernández, D.A., Sánchez-Fuentes, H.J., et al.: Synthesis, characterization and hemolysis studies of $Zn_{(1-x)}Ca_xFe_2O_4$ ferrites synthesized by sol-gel for hyperthermia treatment applications. J. Magn. Magn. Mater. **427**, 241–244 (2017). https://doi.org/10.1016/j.jmmm.2016.10.099
70. Sánchez, J., Cortés-Hernández, D.A., Escobedo-Bocardo, J.C., et al.: Sol-gel synthesis of $Mn_xGa_{1-x}Fe_2O_4$ nanoparticles as candidates for hyperthermia treatment. Ceram. Int. **42**, 13755–13760 (2016). https://doi.org/10.1016/j.ceramint.2016.05.174
71. Mohammadi, H., Nekobahr, E., Akhtari, J., et al.: Synthesis and characterization of magnetite nanoparticles by coprecipitation method coated with biocompatible compounds and evaluation of in-vitro cytotoxicity. Toxicol. Rep. **8**, 331–336 (2021). https://doi.org/10.1016/j.toxrep.2021.01.012
72. Kusigerski, V., Illes, E., Blanusa, J., et al.: Magnetic properties and heating efficacy of magnesium doped magnetite nanoparticles obtained by coprecipitation method. J. Magn. Magn. Mater. **475**, 470–478 (2019). https://doi.org/10.1016/j.jmmm.2018.11.127
73. Klencsár, Z., Ábrahám, A., Szabó, L., et al.: The effect of preparation conditions on magnetite nanoparticles obtained via chemical coprecipitation. Mater. Chem. Phys. (2018). https://doi.org/10.1016/j.matchemphys.2018.10.049
74. Oanh Vuong, T.K., Le, T.T., Do, H.D., et al.: PMAO-assisted thermal decomposition synthesis of high-stability ferrofluid based on magnetite nanoparticles for hyperthermia and MRI applications. Mater. Chem. Phys. **245**, 122762 (2020). https://doi.org/10.1016/j.matchemphys.2020.122762
75. Zargar, T., Kermanpur, A., Labbaf, S., et al.: PEG coated $Zn_{0.3}Fe_{2.7}O_4$ nanoparticles in the presence of <alpha>Fe_2O_3 phase synthesized by citric acid assisted hydrothermal reduction process for magnetic hyperthermia applications. Mater. Chem. Phys. **212**, 432–439 (2018). https://doi.org/10.1016/j.matchemphys.2018.03.054
76. Chin, A.B., Yaacob, I.I.: Synthesis and characterization of magnetic iron oxide nanoparticles via w/o microemulsion and Massart's procedure. J. Mater. Process. Technol. **191**, 235–237 (2007). https://doi.org/10.1016/j.jmatprotec.2007.03.011
77. Zhang, G., Liao, Y., Baker, I.: Surface engineering of core/shell iron/iron oxide nanoparticles from microemulsions for hyperthermia. Mater. Sci. Eng. C **30**, 92–97 (2010). https://doi.org/10.1016/j.msec.2009.09.003

78. Panta, P.C., Romero, R.P.P., Forte, S.K., Bergmann, C.P.: Magnetic synthesis and characterization of superparamagnetic nanoparticles iron oxide stabilized with dextran. In: Ceramic Transactions, pp. 137–145 (2014)
79. Mounkachi, O., Lamouri, R., Abraime, B., et al.: Exploring the magnetic and structural properties of Nd-doped Cobalt nano-ferrite for permanent magnet applications. Ceram. Int. **43**, 14401–14404 (2017). https://doi.org/10.1016/j.ceramint.2017.07.209
80. Milanovic, M., Stijepovic, I., Pavlovic, V., Srdic, V.V.: Functionalization of zinc ferrite nanoparticles: Influence of modification procedure on colloidal stability. Process Appl. Ceram. (2016). https://doi.org/10.2298/PAC1604287M
81. Millot, N., Le Gallet, S., Aymes, D., et al.: Spark plasma sintering of cobalt ferrite nanopowders prepared by coprecipitation and hydrothermal synthesis. J. Eur. Ceram. Soc. (2007). https://doi.org/10.1016/j.jeurceramsoc.2006.04.141
82. Liu, S., Yu, B., Wang, S., et al.: Preparation, surface functionalization and application of Fe_3O_4 magnetic nanoparticles. Adv. Colloid. Interface Sci. **281**, 102165 (2020). https://doi.org/10.1016/j.cis.2020.102165
83. Majidi, S., Zeinali Sehrig, F., Farkhani, S.M., et al.: Current methods for synthesis of magnetic nanoparticles. Artif. Cells, Nanomed. Biotechnol. **44**, 722–734 (2016). https://doi.org/10.3109/21691401.2014.982802
84. Foroughi, F., Hassanzadeh-Tabrizi, S.A., Amighian, J.: Microemulsion synthesis and magnetic properties of hydroxyapatite-encapsulated nano $CoFe_2O_4$. J. Magn. Magn. Mater. **382**, 182–187 (2015). https://doi.org/10.1016/j.jmmm.2015.01.075
85. Rozman, M., Drofenik, M.: Sintering of nanosized MnZn ferrite powders. J. Am. Ceram. Soc. **81**, 1757–1764 (2005). https://doi.org/10.1111/j.1151-2916.1998.tb02545.x
86. Zito, C.A., Orlandi, M.O., Volanti, D.P.: Accelerated microwave-assisted hydrothermal/solvothermal processing: Fundamentals, morphologies, and applications. J. Electroceramics (2018). https://doi.org/10.1007/s10832-018-0128-z
87. Erhardt, C.S., Caldeira, L.E., Venturini, J., et al.: Sucrose as a sol-gel synthesis additive for tuning spinel inversion and improving the magnetic properties of $CoFe_2O_4$ nanoparticles. Ceram. Int. **46**, 12759–12766 (2020). https://doi.org/10.1016/j.ceramint.2020.02.044
88. Venturini, J., Tonelli, A.M., Wermuth, T.B., et al.: Excess of cations in the sol-gel synthesis of cobalt ferrite ($CoFe_2O_4$): A pathway to switching the inversion degree of spinels. J. Magn. Magn. Mater. **482**, 1–8 (2019). https://doi.org/10.1016/j.jmmm.2019.03.057
89. Venturini, J., Zampiva, R.Y.S., Arcaro, S., Bergmann, C.P.: Sol-gel synthesis of substoichiometric cobalt ferrite (CoFe2O4) spinels: Influence of additives on their stoichiometry and magnetic properties. Ceram. Int. **44**, 12381–12388 (2018). https://doi.org/10.1016/j.ceramint.2018.04.026
90. Venturini, J., Wermuth, T.B., Machado, M.C., et al.: The influence of solvent composition in the sol-gel synthesis of cobalt ferrite ($CoFe_2O_4$): A route to tuning its magnetic and mechanical properties. J. Eur. Ceram. Soc. (2019). https://doi.org/10.1016/J.JEURCERAMSOC.2019.01.030
91. Wermuth, T.B., Venturini, J., Guaglianoni, W.C., et al.: Enhancement of magnetic and dielectric properties of $KNbO_3$–$CoFe_2O_4$ multiferroic composites via thermal treatment. Ceram. Int. (2020). https://doi.org/10.1016/j.ceramint.2020.10.060
92. Choudhury, H.A., Choudhary, A., Sivakumar, M., Moholkar, V.S.: Mechanistic investigation of the sonochemical synthesis of zinc ferrite. Ultrason Sonochem **20**, 294–302 (2013). https://doi.org/10.1016/j.ultsonch.2012.06.006
93. Yadav, R.S., Kuřitka, I., Vilcakova, J., et al.: Sonochemical synthesis of Gd^{3+} doped $CoFe_2O_4$ spinel ferrite nanoparticles and its physical properties. Ultrason Sonochem **40**, 773–783 (2018). https://doi.org/10.1016/J.ULTSONCH.2017.08.024
94. Sophia, D., Ragam, M., Arumugam, S.: Synthesis and characterisations of cobalt ferrite nanoparticles. Int. J. Sci. Res. Mod. Educ. 2455–5630 (2016)
95. Gaudisson, T., Vázquez-Victorio, G., Bañobre-López, M. et al.: The Verwey transition in nanostructured magnetite produced by a combination of chimie douce and spark plasma sintering. J. Appl. Phys. Am. Instit. Phys. Inc. 17E117 (2014)

Nanomaterials for Inorganic Pigments

Edgar Andrés Chavarriaga, Tiago Bender Wermuth, and Alex Arbey Lopera

Abstract Inorganic nanopigments have become a field of interest in color generation research. Consequently, the rise of inkjet printing technology has raised the need to obtain pigments with particle sizes that have nanometric values in order to avoid technical problems such as nozzle clogging during printing. Furthermore, inkjet technology requires that inks must remain stable during storage and temperature changes. Inkjet technology has allowed the proposition of new ways of adding color to products that require high processing temperatures, such as ceramic materials. On the other hand, the use of traditional pigments in the final coloring process of a product can be improved at the nanoscale. Therefore, in this chapter, we will review the nanopigments used in inkjet printing technology and in traditional coloring applications in the textile, paper, plastics and ceramics industries.

Keywords Nanopigment · Inkjet printing · Color · Ceramic ink · Coloration

1 Introduction

Inorganic pigments are chemical compounds that, by a process of dispersion of their particles, are used to impart color to another material. A few examples of these materials are wood, concrete, ceramics, plastics, cosmetics, textiles and food, which are used either for a decorative purpose, e.g. in architectural paintings, or to fulfill some specific function according to an intrinsic property of the compound, e.g. anti-corrosive, luminescent, magnetic, electrochromic and photochromic properties. Due

E. A. Chavarriaga (✉)
Grupo GICEI, Institución Universitaria Pascual Bravo, Calle 73 No 73A -226, Medellín, 050034, Colombia
e-mail: e.chavarriaga2556@pascualbravo.edu.co

T. B. Wermuth
Programa de Pós-Graduação Em Engenharia de Minas, Metalúrgica E de Materiais, Universidade Federal Do Rio Grande Do Sul, Porto Alegre, Brazil

A. A. Lopera
Universidad Nacional de Colombia Sede de La Paz, La Paz, Dirección académica Km 9 vía Valledupar La Paz, La Paz, 202010, Valledupar, Colombia

A. Kopp Alves (ed.), *Technological Applications of Nanomaterials*,
Engineering Materials, https://doi.org/10.1007/978-3-030-86901-4_11

to the processing conditions of the materials that the pigments color—for instance, vitreous enamels or ceramic bodies in the ceramic industry–, they must remain stable at high temperatures and also have high chemical resistance to the components with which they interact.

Since ancient times, the fascination of man with the color of nature has prompted him to look for the substances responsible for color. Initially, only natural ones were explored, in processes that required only their manual extraction from the site where they were found, as evidenced by the archaeological findings of objects manufactured since prehistory. With the passage of time, however, the discovery of fire allowed the manipulation of substances at high temperatures and therefore the development of new inorganic pigments. Among these, it is interesting to mention Egyptian blue ($CaCuSi_4O_{10}$), which owes its color to the Cu^{2+} cation when it is in planar square coordination. A remarkable fact about this long walk through history is that new inorganic pigments are still being sought today, as substitutes for traditional ones due to the effects that some components of the latter have on humans, soil, water or air.

In recent years, research on inorganic nanopigments has been promoted in order to explore new applications of these substances [1–4]. Among the nanometric particles used as pigments, it is important to highlight the following: TiO_2 [5], $CoAl_2O_4$ [6], Fe_2O_3/TiO_2 [7], $Ca_3Cr_2Si_3O_{12}$ [8], $Sr_2SiO_4{:}Sm^{3+}$ [9], $BiVO_4$ [10], $SrAl_2O_4{:}Eu^{2+}/Dy^{3+}$ [11], Cr-doped ZnO [3], Fe-doped Al_2O_3 [12], V-ZrO_2 [13], $Zn_2P_2O_7$ and $Zn_3(PO_4)_2$ [14], $Ce_xZr_{1-x}O_2$ [15], $Cr_{1.3}Fe_{0.7}O_3$ [16], Bi-doped ZnO[17], $CoFe_2O_4$ [18], $Cr_xSb_xTi_{1-2x}O_2$ [19], $CoCr_2O_4$[20], $ZnCo_xAl_{2-x}O_4$ [21], Ga-doped ZnO [22], $CoCr_{2-x}Al_xO_4$ [23], $PbCrO_4$ [24], and $MnAl_2O_4$ [25].

1.1 Nanopigments for Inkjet Printing

Inorganic nanopigments with chromophoric ions have been prepared mainly to be applied using ink-jet technology, which requires particle sizes of less than 200 nm in order to avoid the blockage of the inkjet printer nozzle. Heydari et al. [26] studied the effect of solvents with glycol structures, as well as cetyltrimethylammonium bromide and polyvinylpyrrolidone, as dispersants to form stable suspensions of the blue pigment V-$ZrSiO_4$ using the polyol method. The powders obtained had particle sizes ranging from 10 to 50 nm, which showed that the length of the hydrocarbon chain of the polyol affects the particle size of the suspension. The diethylene glycol used in the experiment formed stable suspensions for 2 months, then the resulting ink was applied to ceramic tiles and calcined at 1000 °C, generating a blue color. In another study, nanoparticles of a $CoAl_2O_4/Al_2O_3$ composite system were synthesized using incipient wetness impregnation, as shown in Fig. 1, to develop pigments for the preparation of inks to be used in inkjet printing [27].

The exploration of substances as nanopigments for inkjet applications has not been done using inorganic oxides only. As shown in Fig. 2, Blosi et al. [28] obtained

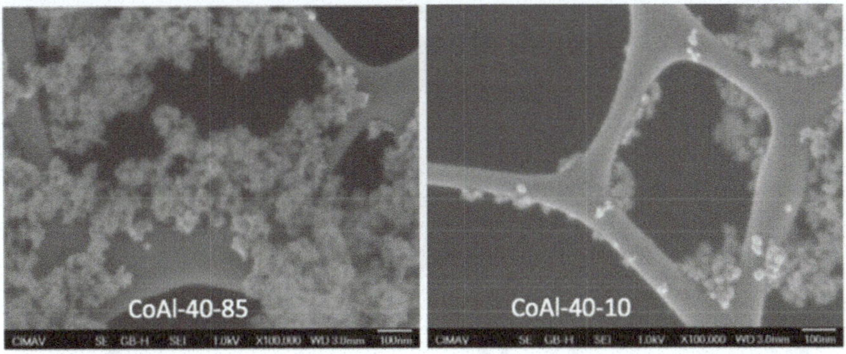

Fig. 1 SEM micrographs of blue pigments with nanometric sizes. Reproduced from Ref. [27] with permission from Elsevier

Fig. 2 Color shift in ceramic inks with different Ag/Au ratios. Reproduced from Ref. [28] with permission from Elsevier

Au and Ag nanosuspensions to be used in ceramic tile decoration using microwave-assisted synthesis. Better results were acquired for the pieces calcined at 1150 °C when a 0.04% Au concentration, an 1.5 Au/Ag ratio and a particle size of less than 12 nm were used.

$CoAl_2O_4$, a blue inorganic nanopigment with a spinel structure, was synthesized by Peymannia et al. [29] using the coprecipitation synthesis method. The authors found that its particle size is between 40 and 55 nm. In their research, inks were prepared and applied to ceramic substrates using an inkjet printer, and the resulting samples were calcined at 1100 °C. The final pieces are shown in Fig. 3, where A, B and C mean the number of runs in the printing process: 1, 3 and 5 runs, respectively. The results show that the blue color is more intense when the thickness of the printed film increases.

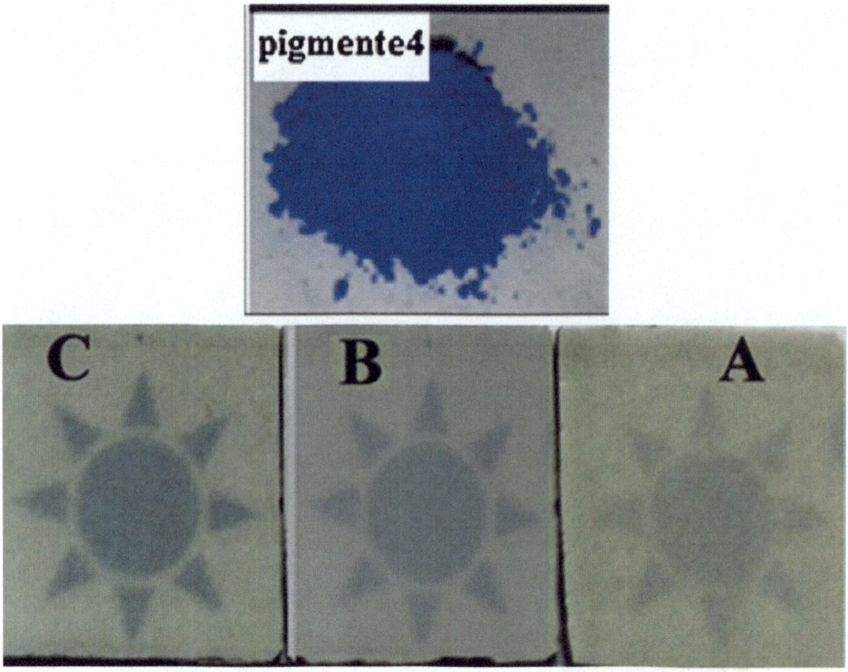

Fig. 3 $CoAl_2O_4$ nanopigment and ceramic pieces printed using ink. Reproduced from Ref. [29] with permission from Elsevier

One of the fields of research on the production of nanopigments has focused on using synthesis pathways that have an easy experimental setup and allow powders to be obtained in a short time. Adopting a similar methodology, Tan et al. [30] synthesized $CoAl_2O_4$ doped with Cu^{2+} using solution combustion synthesis, which involves promoting the oxidation reaction of a fuel using nitrates as oxidants. This process releases heat and allows the necessary temperatures for the formation of the desired composition to be reached, as shown in Fig. 4. The synthesized $Cu_{0.1}Co_{0.9}Al_2O_4$ pigments have an approximate particle size of 30 nm and were applied as ceramic ink at 1200 °C, showing good blue color stability.

In another study, Zhao et al. [31] used threonine as a complexing agent in the synthesis of $CoAl_2O_4$. The authors employed the coprecipitation synthesis method to avoid particle agglomeration during the calcination process at 1000° C for 2 h, and they found that the particle size of the powders was between 10 and 20 nm. Following that, they made two inks with threonine and without threonine as shown in Fig. 5, which demonstrated that threonine allows the formation of smaller particles.

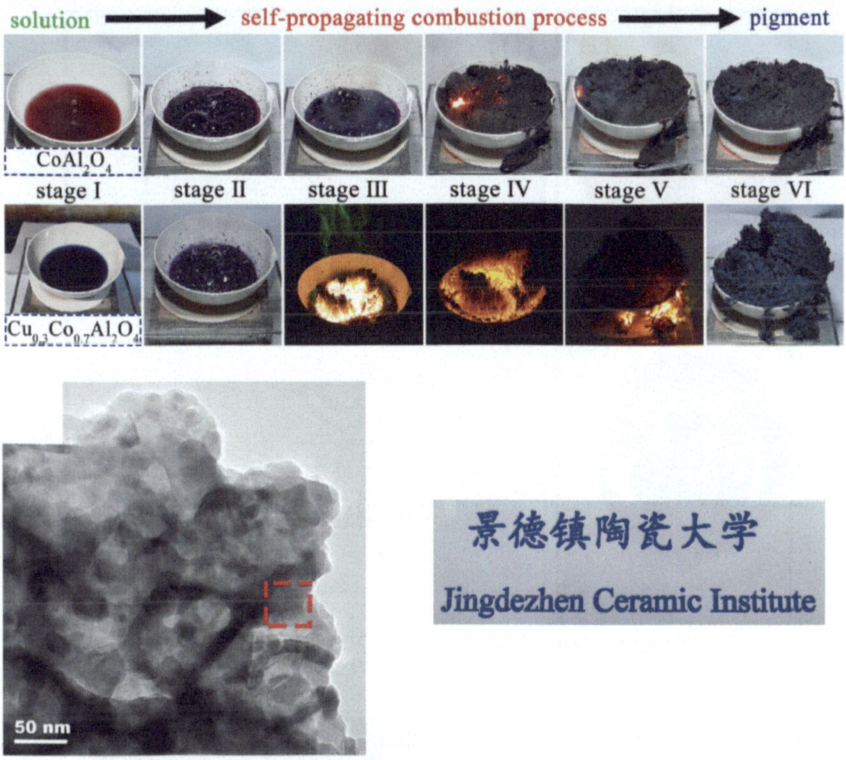

Fig. 4 Digital images of the combustion synthesis process, TEM analysis and application as ceramic ink of $Cu_{0.1}Co_{0.9}Al_2O_4$. Reproduced from Ref. [30] with permission from Elsevier

1.1.1 Nanopigments in Other Applications

Nanopigments allow the colorimetric properties of powders to be changed, since the dispersion of the latter depends on their size. For that reason, alternative synthesis pathways to traditional methods have recently been proposed in which the size of the particles that make up the pigments is controlled. At the nanoscale, contrary to the case of inkjet printing, their properties can be changed in order for them to be used in other fields, such as textiles, plastics, cement, paper and ceramics.

For example, $Y_2Cu_2O_5$, a blue-green nanopigment with a particle size of 47 nm, was synthesized using solution combustion synthesis [32]. In this case, it is worth noting that when the particle size of the pigments decreases, the diffuse reflectance spectrum increases. When compared with the powder synthesized using the solid state reaction method—which was created due to an increase in the dispersion of electromagnetic radiation as a consequence of the change in particle size, as observed in Fig. 6, the pigment remained stable after it was applied to ceramic glazes. On the other hand, white zinc molybdate and zinc phosphomolybdate nanopigments with anticorrosive properties were obtained using the ultrasound method. They have

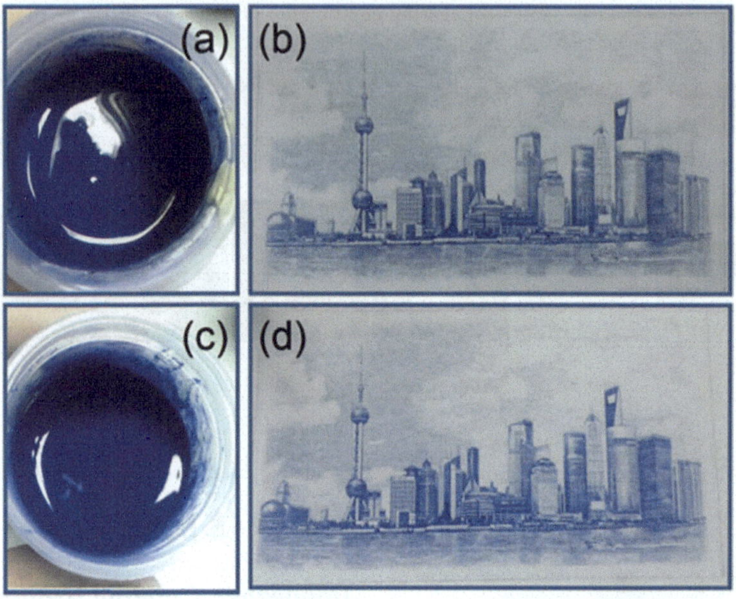

Fig. 5 Digital photos of inks and printings using the $CoAl_2O_4$ nanopigment **a**, **b** without threonine and **c**, **d** with threonine. Reproduced from Ref. [31] with permission from Elsevier

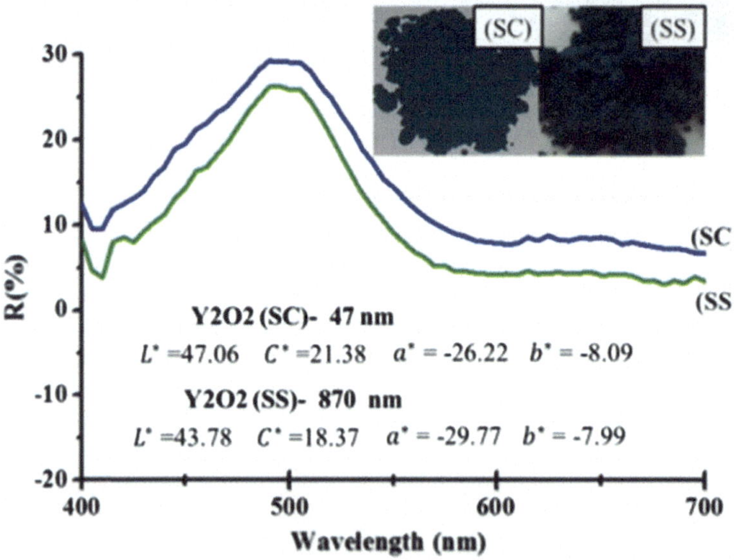

Fig. 6 Reflectance spectra for the $Y_2Cu_2O_5$ nanopigments synthesized by the sol–gel-combustion (SC) and solid state reaction (SS) methods. Reproduced from Ref. [32] with permission from Elsevier

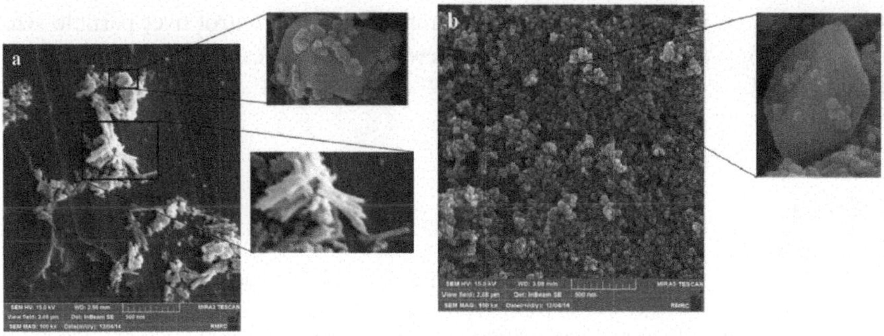

Fig. 7 Nano iron oxide morphology: **a** Fe_3O_4/urea and **b** Fe_3O_4/TEA. Reproduced from Ref. [33] with permission from Elsevier

particle sizes ranging from 21 nm for $ZnMoO_4$ to 95.2 nm for Zn_2PMoO_7, which shows that these nanopigments can be synthesized by the alternate method more rapidly in comparison to the conventional synthesis method, i.e. coprecipitation.

Recently, Javidpavar et al. [33] studied the effect that different nanoparticle morphologies of the inorganic pigment Fe_3O_4 have on the mechanical properties of the manufacturing process of an epoxy coating. Nanoparticles were synthesized using a precipitation process with the surfactants triethanolamine (TEA) and urea, in addition to ferrous and ferric iron chlorides as precursors, and NaOH as a precipitating agent, in a reducing atmosphere composed of N_2.

The crystallite sizes of the powders were estimated using the Scherrer equation and it was found that they are 18 nm for Fe_3O_4/TEA and 39 nm for Fe_3O_4/urea. The morphology of the particles can change from fibers in urea, to polygons in TEA (Fig. 7). When the resulting nanopigment had a fibrous morphology and was used in the epoxy coating, the mechanical properties of the material were improved due to the orientation of the fibers. Additionally, Piri et al. [4] analyzed the influence of a TiO_2 nanopigment on the optical properties of a polyester fabric used in the textile industry. They found that particles with a size of 35 nm increased UV absorption.

2 Conclusion

Inorganic nanopigments show that new applications in materials engineering can be obtained, since these substances allow the changing of the physical properties that are relevant to technology. The optical properties of inorganic nanopigments are their most studied features, but it is still important to continue accumulating knowledge about the influence that nanoparticles have on the systems to which they are applied, whether it be inkjet printing technology or ceramic decoration. Other important subjects that must be further studied are the interaction of pigments with other compounds that integrate the substrates with which they interact, the exploration of new synthesis methods that allow researchers to have control over particle size and morphology, and finally the study of new nanostructures as potential inorganic pigments.

References

1. El-Sherbiny, S., Morsy, F., Samir, M., Fouad, O.A.: Synthesis, characterization and application of TiO_2 nanopowders as special paper coating pigment. Appl. Nanosci. **4**, 305–313 (2014). https://doi.org/10.1007/s13204-013-0196-y
2. Laguna, M., Núñez, N.O., Fernández, M., Ocaña, M.: Synthesis and optical properties of environmentally benign and highly uniform $NaCe(MoO_4)_2$ based yellow nanopigments. J. Alloys Compd. **739**, 542–548 (2018). https://doi.org/10.1016/j.jallcom.2017.12.158
3. ALkallas, F.H., Elshokrofy, K.M., Mansour, S.A.: Structural and diffuse reflectance study of Cr-doped ZnO nanorod-pigments prepared via facile thermal decomposition technique. J. Inorg. Organomet. Polym. Mater. **29**, 792–798 (2019). https://doi.org/10.1007/s10904-018-01053-4
4. Piri, N., Shams-nateri, A., Mokhtari, J.: Solar spectral performance of nanopigments. Sol. Energy Mater. Sol. Cells. **162**, 72–82 (2017). https://doi.org/10.1016/j.solmat.2016.12.036
5. Shams-Nateri, A., Kazemian, S., Piri, N.: Nano-TiO_2 coated cotton fabrics with temperature regulating properties. J. Text. Inst. **111**, 1223–1230 (2020). https://doi.org/10.1080/00405000.2019.1690916
6. Tatarchuk, T., Shyichuk, A., Lamkiewicz, J., Kowalik, J.: Inversion degree, morphology and colorimetric parameters of cobalt aluminate nanopigments depending on reductant type in solution combustion synthesis. Ceram. Int. **46**, 14674–14685 (2020). https://doi.org/10.1016/j.ceramint.2020.02.269
7. Lee, H.J., Jung, K.Y., Kim, Y.-S.: Nanostructured Fe_2O_3/TiO_2 composite particles with enhanced NIR reflectance for application to LiDAR detectable cool pigments. RSC Adv. **11**, 16834–16840 (2021). https://doi.org/10.1039/d1ra02614c
8. Rendón-Angeles, J.C., Matamoros-Veloza, Z., Rodríguez-Galicia, J.L., Seong, G., Yanagisawa, K., Tamayo, A., Rubio, J., Anaya-Chavira, L.A.: One-pot hydrothermal synthesis of victoria green ($Ca_3Cr_2Si_3O_{12}$) nanoparticles in Alkaline fluids and its colour hue characterisation. Nanomaterials **11**, 1–15 (2021). https://doi.org/10.3390/nano11020521
9. Manjunath, C., Rudresha, M.S., Walsh, B.M., Hari Krishna, R., Panigrahi, B.S., Nagabhushana, B.M.: Optical absorption intensity analysis using Judd-Ofelt theory and photoluminescence investigation of orange-red Sr_2SiO_4: Sm^{3+} nanopigments. Dye. Pigment. **148**, 118–129 (2018). https://doi.org/10.1016/j.dyepig.2017.08.036
10. Patil, V.J., Bhoge, Y.E., Patil, U.D., Deshpande, T.D., Kulkarni, R.D.: Room temperature solution spray synthesis of Bismuth Vanadate nanopigment and its utilization in formulation of industrial OEM coatings. Vacuum **127**, 17–21 (2016). https://doi.org/10.1016/j.vacuum.2016.02.005
11. Abumelha, H.M.: Simple production of photoluminescent polyester coating using lanthanide-doped pigment. Luminescence **36**, 1024–1031 (2021). https://doi.org/10.1002/bio.4030
12. Heiba, Z.K., Mohamed, M.B., Wahba, A.M., Imam, N.G.: Structural, Optical, and Electronic Characterization of Fe-Doped Alumina Nanoparticles. J. Electron. Mater. **47**, 711–720 (2018). https://doi.org/10.1007/s11664-017-5830-0
13. Calatayud, J.M., Alarcón, J.: V-containing ZrO_2 inorganic yellow nano-pigments prepared by hydrothermal approach. Dye. Pigment. **146**, 178–188 (2017). https://doi.org/10.1016/j.dyepig.2017.07.003
14. Alibakhshi, E., Naeimi, A., Ramezanzadeh, M., Ramezanzadeh, B., Mahdavian, M.: A facile synthesis method of an effective anti-corrosion nanopigment based on zinc polyphosphate through microwaves assisted combustion method; comparing the influence of nanopigment

and conventional zinc phosphate on the anti-corrosion properties of an e. J. Alloys Compd. **762**, 730–744 (2018). https://doi.org/10.1016/j.jallcom.2018.05.172
15. Joseph, A., John Mathew, K.P., Vandana, S.: Zirconium-doped ceria nanoparticles as anticorrosion pigments in waterborne epoxy-polymer coatings. ACS Appl. Nano Mater. **4**, 834–849 (2021). https://doi.org/10.1021/acsanm.0c03162
16. Shaterian, M., Khoobi, A., Enhessari, M., Ozaee, K.: A new strategy based on thermodiffusion of ceramic nanopigments into metal surfaces and formation of anti-corrosion coatings. Microporous Mesoporous Mater. **218**, 62–68 (2015). https://doi.org/10.1016/j.micromeso.2015.06.039
17. Mastan, R., Khorsand Zak, A., Pilevar Shahri, R.: Bi-doped ZnO yellow nanopigments: Synthesis, characterization, and antibacterial application for painting humid places. Ceram. Int. **46**, 8582–8587 (2020). https://doi.org/10.1016/j.ceramint.2019.12.090
18. Basak, M., Rahman, M.L., Ahmed, M.F., Biswas, B., Sharmin, N.: Calcination effect on structural, morphological and magnetic properties of nano-sized $CoFe_2O_4$ developed by a simple co-precipitation technique. Mater. Chem. Phys. **264**, 124442 (2021). https://doi.org/10.1016/j.matchemphys.2021.124442
19. Calatayud, J.M., Pardo, P., Alarcón, J.: Cr- and Sb-containing TiO_2 inorganic orange nanopigments prepared by a relative long hydrocarbon chain polyol. Dye. Pigment. **134**, 1–8 (2016). https://doi.org/10.1016/j.dyepig.2016.06.034
20. Chamyani, S., Salehirad, A., Oroujzadeh, N., Fateh, D.S.: Effect of fuel type on structural and physicochemical properties of solution combustion synthesized $CoCr_2O_4$ ceramic pigment nanoparticles. Ceram. Int. **44**, 7754–7760 (2018). https://doi.org/10.1016/j.ceramint.2018.01.205
21. Wahba, A.M., Imam, N.G., Mohamed, M.B.: Flower-like morphology of blue and greenish-gray $ZnCo_xAl_{2-x}O_4$ nanopigments. J. Mol. Struct. **1105**, 61–69 (2016). https://doi.org/10.1016/j.molstruc.2015.10.052
22. Farha, A.H., Ibrahim, M.M., Mansour, S.A.: Ga-doped ZnO nanostructured powder for cool-nanopigment in environment applications. Materials (Basel). **13**, 1–16 (2020). https://doi.org/10.3390/ma13225152
23. El Jabbar, Y., Lakhlifi, H., El Ouatib, R., Er-Rakho, L., Guillemet-Fritsch, S., Durand, B.: Green nanoceramic pigments based on cobalt chromite doped with Al: Synthesis, characterisation and colour performance. Ceram. Int. **47**, 9373–9381 (2021). https://doi.org/10.1016/j.ceramint.2020.12.068
24. Deshpande, P.S., Patil, V.J., Mahulikar, P.P., Patil, U.D., Kulkarni, R.D.: Interfacial analysis and reaction engineering of sucrose ester mediated solution spray synthesis of lead chromate nanorods. Chem. Eng. Process. Process Intensif. **95**, 390–402 (2015). https://doi.org/10.1016/j.cep.2015.07.026
25. Ghasemi, E., Ramezanzadeh, B., Saket, S., Ashhari, S.: Electrochemical investigation of the epoxy nanocomposites containing $MnAl_2O_4$ and $CoAl_2O_4$ nanopigments applied on the aluminum alloy 1050. J. Coatings Technol. Res. **13**, 97–114 (2016). https://doi.org/10.1007/s11998-015-9728-6
26. Heydari, H., Yousefpour, M., Emadoddin, E., Zori, M.H., Aminian, M.K.: Effect of solvents and dispersants on polyol synthesis of V-$ZrSiO_4$ nanopigment stable suspension for ink application. J. Coat. Technol. Res. **17**, 1243–1253 (2020). https://doi.org/10.1007/s11998-020-00343-2
27. Aguilar-Elguézabal, A., Román-Aguirre, M., De la Torre-Sáenz, L., Pizá-Ruiz, P., Bocanegra-Bernal, M.: Synthesis of $CoAl_2O_4/Al_2O_3$ nanoparticles for ceramic blue pigments. Ceram. Int. **43**, 15254–15257 (2017). https://doi.org/10.1016/j.ceramint.2017.08.062
28. Blosi, M., Albonetti, S., Gatti, F., Baldi, G., Dondi, M.: Au–Ag nanoparticles as red pigment in ceramic inks for digital decoration. Dye. Pigment. **94**, 355–362 (2012). https://doi.org/10.1016/j.dyepig.2012.01.006
29. Peymannia, M., Soleimani-Gorgani, A., Ghahari, M., Najafi, F.: Production of a stable and homogeneous colloid dispersion of nano $CoAl_2O_4$ pigment for ceramic ink-jet ink. J. Eur. Ceram. Soc. **34**, 3119–3126 (2014). https://doi.org/10.1016/j.jeurceramsoc.2014.03.022

30. Tan, L., Wang, Q., Cheng, Z., Hu, Z., Liu, K., Wang, Y., Chang, Q.: Research on the low-temperature synthesis of cobalt aluminum spinel type blue pigments. J. Alloys Compd. **864**, 158625 (2021). https://doi.org/10.1016/j.jallcom.2021.158625
31. Zhao, S., Guo, J., Li, W., Guo, H., You, B.: Fabrication of cobalt aluminate nanopigments by coprecipitation method in threonine waterborne solution. Dye. Pigment. **151**, 130–139 (2018). https://doi.org/10.1016/j.dyepig.2017.12.062
32. Farbod, M., Rafati, Z.: Color parameters of $Y_2Cu_2O_5$ green-blue nanopigments fabricated by the sol-gel combustion method and their efficiency for coloring the glazed tiles. Ceram. Int. **42**, 15732–15738 (2016). https://doi.org/10.1016/j.ceramint.2016.07.033
33. Javidparvar, A.A., Ramezanzadeh, B., Ghasemi, E.: Effects of surface morphology and treatment of iron oxide nanoparticles on the mechanical properties of an epoxy coating. Prog. Org. Coatings. **90**, 10–20 (2016). https://doi.org/10.1016/j.porgcoat.2015.09.018

Graphene Application

Clívia Kellen Almeida Silva and Annelise Kopp Alves

Abstract Graphene is an allotropic form of carbon with hexagonal structures with highly sought-after chemical and physical properties. This chapter initially presents a review of methods to obtain and characterizing graphene and a literature summary of applications of this material. In the end, expectations for future work are presented.

Keywords Graphene · Characterization · Nanostructure

Abbreviations

XRD	X-ray diffraction
rGO	Reduce graphene oxide
LIBs	Lithium-ion batteries
GNR	Graphene nano-ribbon
GO	Graphene oxide
QD	Graphene quantum
CQDs	Carbon quantum dot
GQDs	Graphene quantum dot
HRTEM	High resolution transmition electron microscopy
SEM	Scanning electron microscopy
TEM	Transmition electron microscopy
EELS	Electron energy loss spectroscopy
EDX	Energy dispersive x-ray
LED	Light emitting diode
WLED	White light emitting diode
PEC	Hidrogenio fotoeletronico
UHV	Ultrahigh vacuum

C. K. A. Silva (✉) · A. K. Alves
Universidade Federal Do Rio Grande Do Sul, Porto Alegre, Brazil

A. Kopp Alves (ed.), *Technological Applications of Nanomaterials*,
Engineering Materials, https://doi.org/10.1007/978-3-030-86901-4_12

1 Introduction

Graphene is a carbon allotrope whose applications have grown significantly in the last decade. As a result, many works have been published addressing these applications, characterizations, and synthesis methods of this bidimensional nanostructure.

Graphene has a hexagonal structure and the smallest thickness that a material can have, less than 1 nm. In its structure, there are carbons with sp^2 hybridization in which each carbon is linked with three other carbons by σ bonds [1]. Figure 1 [2] shows a representation of carbon atoms in a graphene lattice. Figure 1a, a ball-and-stick model, and Fig. 1b, a model with filled spaces is represented. The arrows in all directions indicate the infinite extension of their ends.

The physical properties that graphene exhibits are what make it so attractive. Its existence was nothing new; what was puzzling was how it would be extracted from graphite. In 2004, two researchers from the university of manchester, Professor Andre Geim and Professor Kostya Novoselov, won the Nobel Prize in Physics in 2010 [3] for this development.

Could it be possible to imagine that graphene would be discovered on a Friday using adhesive tape and graphite fragments?

This is how the work on attaining graphene was started: by repeated separations, a flake as thick as one atom was obtained, and thus the first isolation of graphene was achieved. Nowadays, the need for research on synthesis and characterization methods so that the applications of graphene remain viable is of paramount interest. Moreover, because graphene has a wide range of properties, the required properties must focus on synthesizing the composite.

The nature of electrical charge portability of graphene is unusual; it reaches a speed of 300 times less than that of light which grants it the property of excellent electrical conductivity. As for the thermal properties of graphene, these are guaranteed due to its covalent bonds between the carbons. These two properties make graphene a material with enormous potential for use in electronic devices [1]. Furthermore, graphene

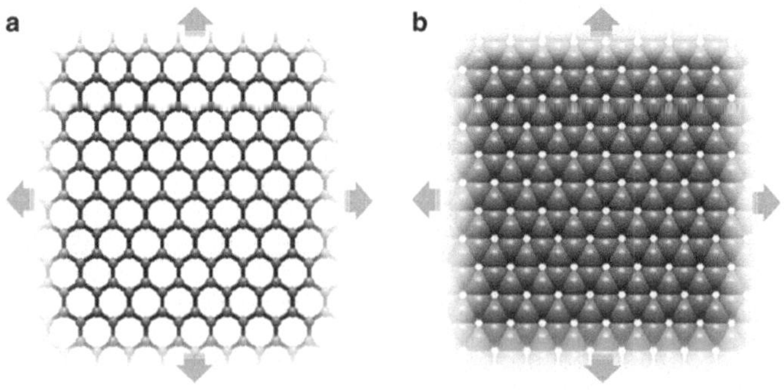

Fig. 1 Schematic representation of a graphene lattice. Reprinted with permission from Kar et al. [2]

has one π bond available on each carbon atom, and it is this electron guarantees its electrical and optical properties [4].

Materials with carbon in the composition, such as diamond and graphite, have high hardness and Young's ratio values, so one would expect the same for graphene. However, their nanometer measurements make proper testing difficult. Nevertheless, through mathematical models, it was possible to estimate high values for such properties [1].

Graphene can be applied in several areas such as biomedical, composites and coatings, electronics, and energy areas, all due to its properties of strength, lightness, flexibility, and conductivity. Would it be possible to imagine bending a cell phone or tablet? Or to have a paint that simply inhibits rust on iron? To have batteries that last much longer and consume much less energy during the charging time and yet are light in such a way that the weight is not a limiting factor in their use? Or to solve the problem of water shortage in arid and dry areas? All is possible with graphene technology. But the commercialization of graphene presents a challenge: one is producing this nanomaterial with quality on a large scale and the other, do it at a low cost [5]. That is why exploring and choosing the best characterization and synthesis technique is essential according to the desired property.

In this chapter, you will learn how graphene is obtained, characterized, and some examples of applications.

2 Ways of Obtaining Graphene

The most suitable way to obtain graphene will depend a lot on the scale to be produced.

Graphene is obtained from graphite, and the way it was discovered by exfoliation with adhesive tape is still used; this technique is called micromechanical exfoliation, but it is not a viable technique for large-scale production is more commonly used as a research application.

2.1 Chemical Vapor Phase Deposition (CVD)

In this method, graphene grows on metal surfaces of Ni, Co, or Cu. This substrate is placed in a CVD chamber under vacuum conditions at 10^{-3} Torr with dilute hydrocarbon gases at a temperature below 1000 °C. Thus, the carbon adheres to the metal surface after cooling down, and through a chemical attack on the metal substrate, the graphene layers are transferred to another surface [1].

2.2 Epitaxial Growth on SiC

Under UHV conditions, a SiC (6H-SiC) substrate is heated where the Si atoms sublimate, and the carbon atoms stay on the surface and form graphene. Conditions such as temperature and annealing time directly influence the number of graphene layers that will be obtained. Therefore, a short annealing time will result in few layers of graphene. Other factors that can also influence are the use of Ar, Ir, Ni, Co, and Pt during the synthesis [1].

This method is attractive to the semiconductor industries, as the graphene is already adhered to the Si surface, avoiding the movement of graphene to another surface. With a remark that the lack of control of the graphene layers may occur, impairing the thermal and electrical properties [1].

2.3 Liquid-Phase Chemical Exfoliation

Graphite has staking π-π interactions, which are intermolecular stacking interactions between the graphene layers [6]. Promoting chemical exfoliation of liquid phase graphite makes it possible to decrease and break these interactions through reagents such as potassium chloride and nitric and sulfuric acids [7].

The graphite is placed in contact with organic solvents or aqueous solutions to be exfoliate when exposed to an energy source, usually ultrasonic waves, for a period of time. This way, polydisperse graphene with varying numbers of layers is obtained. Centrifugation is required to remove unexfoliated graphite particles and allow the separation of graphene layers [8]. The great advantage of this method is that it produces graphene on a large scale.

3 Characterization Methods

3.1 X-Ray Diffraction (XRD)

X-Ray Diffraction (XRD) is a technique used to determine the structure of a material when an X-ray beam reaches into its structure. The material's lattice diffracts the X-rays, and a collector analyses the position and intensity of the diffracted rays, forming a diffractogram.

Figure 2 shows examples of the characteristic X-ray diffractograms of graphite, graphene oxide (GO) obtained with the Hummer method, and graphene oxide in ionic liquid mixtures (GO-IL) prepared by other methods [9].

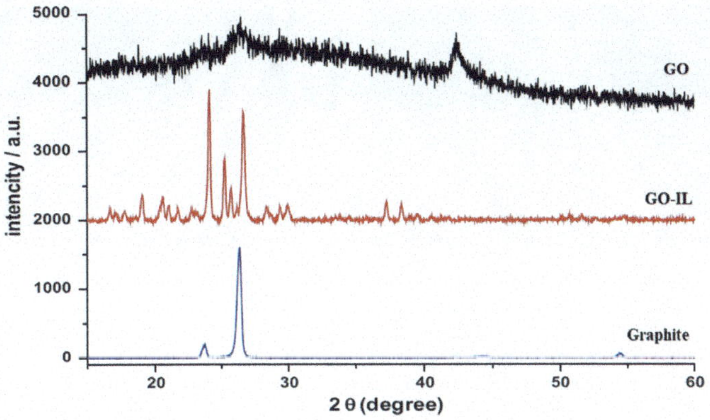

Fig. 2 DRX spectrum of graphite, graphene oxide (GO), and graphene oxide in an ionic liquid mixture (GO-IL). Adapted and reprinted with permission of Kadari et al. [9]

3.2 *Scanning Electron Microscopy (SEM)*

The scanning electron microscopy (SEM) uses an electron beam's incidence on the surface of a structure to create a detailed image. In this interaction, signals are emitted, and information on the material's surface morphology and chemical composition can be obtained.

An excellent example of the use of SEM for the characterization of graphene is presented in the work of Coroş et al. [5]. The author prepares samples by electrochemical exfoliation from graphite anode in a potassium sulfate solution. By varying electrochemical parameters, a study of the dimensions of graphene flakes and the level of exfoliation/oxidation was performed. The electrochemical parameters considered were polarization and electrolyte concentration.

Figure 3 shows the SEM images of the GO structures obtained in the work of Azizighannad et al. [10], where graphene was subjected to mild chemical reduction.

3.3 *Transmission Electron Microscopy (TEM)*

Similar to SEM, TEM uses an electron beam that forms the image of the sample. Electrons move in a tube in a vacuum, and the beam passes through the sample, and an image is formed on a screen.

Figure 4 from Shaikh et al. [11] presents TEM images of the graphene layers obtained after the electrochemical exfoliation.

Figure 4a, b, c, d detail the graphene layers obtained by exfoliation in different conditions. In Fig. 4a, graphene sheets were obtained without added H_2O_2, while Figs. 4b, c present images of the aspect of the layers obtained with added H_2O_2, 5 and 10 mL, respectively, at 50 °C. In Fig. 4d, graphene layers were obtained after

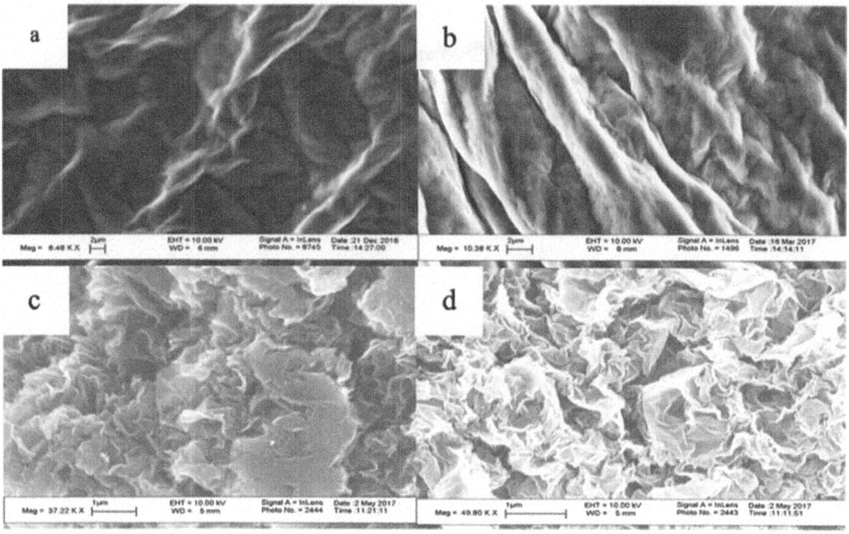

Fig. 3 SEM images of **a** GO, **b** r-GO-31, **c** r-GO-19, **d** r-GO-9. Reproduced with permission from Azizighannad et al. [10]

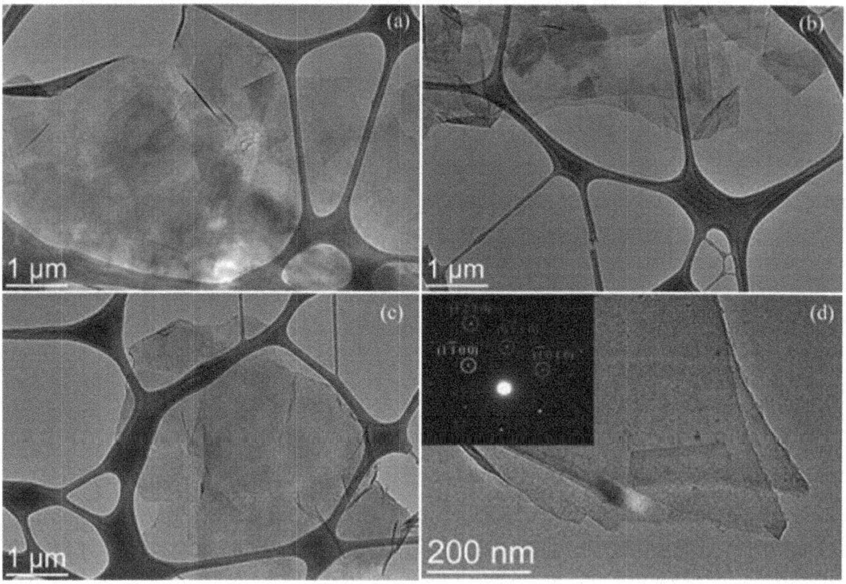

Fig. 4 TEM images of the exfoliated graphene synthesized using electrochemical exfoliation at 50 °C. **a** without the addition of H_2O_2, **b** with the addition of 5 mL H_2O_2, and **c** 10 mL H_2O_2. **d** Electrochemical exfoliation at 95 °C with the addition of 10 mL H_2O_2 and the corresponding selected area electron diffraction pattern (inserted). Reproduced with permission from Shaikh et al. [11]

chemical exfoliation at 95 °C using 10 mL of H_2O_2. The temperature increase and the presence of H_2O_2 help to exfoliate graphite more efficiently.

3.4 *Raman Spectroscopy*

Raman spectroscopy is used to obtain chemical information and material structure through a photonic technique [12].

The Raman spectrum for carbon-related materials inform particularly three bands:

- D-band in the range of 1250–1350 cm^{-1}, associated with defects/disordered structure of graphene, is a measure of the sp^3 bonding in graphene. The absence of this band in the Raman spectrum indicates that the materials are free from defects;
- G-band represents the sp^2 bonded carbon that constitutes graphene. The position and the shape of the G-band peak provides information about the thickness of the graphene;
- G′-band, also known as the 2D band, appears in the absence of defects or impurities. The G′-band is the result of a two phonon lattice vibrational process.

In the work of Shaikh et al. [11], Raman spectroscopy was used to detect the effect of exfoliation conditions (presence of H_2O_2 and temperature) in the quality of the obtained graphene. In Fig. 5 the Raman spectra shows three characteristic peaks: D (1325 cm^{-1}), G (1568 cm^{-1}) and 2D (2376 cm^{-1}).

A small peak at 1325 cm^{-1} (D band) shows the high purity of graphene. It can be seen that the intensity of D and G decrease with increasing concentration of the electrolyte solution, which is an indication of better quality graphene.

4 Applications

In the search for applications of graphene, a comprehensive bibliography was found in its uses in several areas with a greater concentration on composites. Properties such as increased conductivity, mechanical strength, and thermal properties due to a small amount in polymers and increased conductivity through forming an electron transport layer in solar cells [12].

4.1 *Applications of Graphene in Improvising Conductivity*

As the years progress, energy resources become a challenge for new sources or improvements to existing sources. For lithium-ion batteries (LIBs), the energy storage capacity depends on the material that makes up the electrodes and their properties. Several authors investigated the applicability and performance of graphene

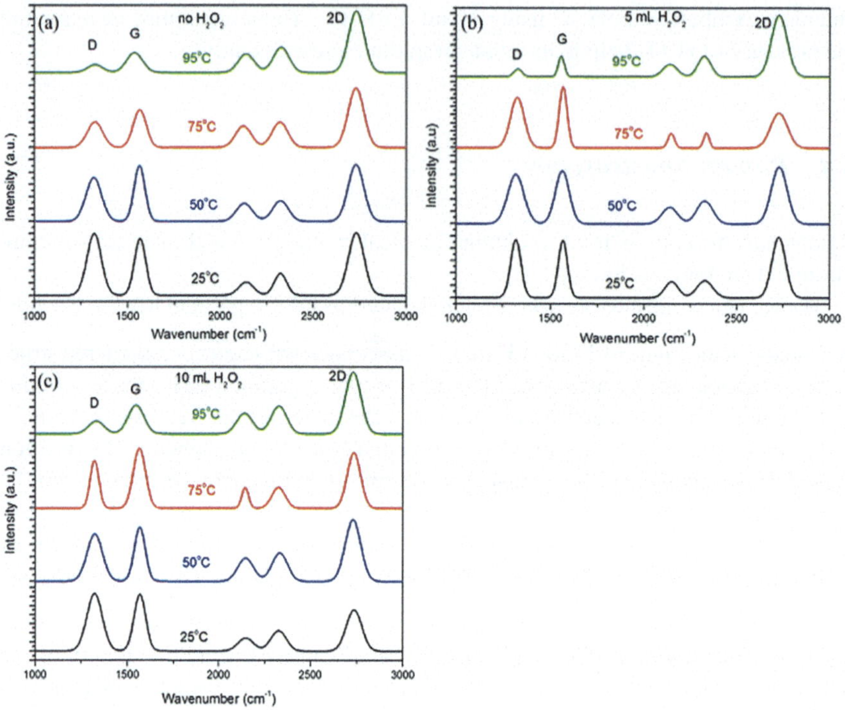

Fig. 5 Raman spectra of graphenes synthesized using electrochemical exfoliation. **a** without the addition of H_2O_2, **b** with the addition of 5 mL H_2O_2, and **c** with 10 mL H_2O_2, at different temperatures. Reproduced with permission from Shaikh [11]

and reduced graphene oxide (rGO), as it has a higher performance, for electrode composites in lithium batteries were investigated.

In the work of Lu et al. [13] on applications of rGO in lithium-ion batteries, graphene, and related materials were used as an additive in the anode and cathode of these batteries. Due to the high conductivity of graphene and its mechanical properties, improvements in LIBs were observed.

In the same field of research, graphene hybrids were also presented as a possible enhancer for LIBs. One of them is based on a 3D and flexible porous graphene without a binder [14, 15].

Zhou et al. [14] studied Si nanoparticles' dispersion through electrostatic action between one layer and another of graphene that presented a superior conductivity performance.

In the work of Yi et al. [15], Si was also studied for use in a composite with graphene. In this work, the high energy capacity in relation to other commercial anodes is explained due to the low electrical resistance at high mass loads.

According to some researchers in recent years, to improve the performance of electrodes composed of graphene, it is necessary to obtain an ideal morphological design [13]. Thus, 3D porous structures are designed with a high pore volume to make a total utilization surface area possible [16].

It can be summed up that graphene is a material with the potential to solve the problems faced by lithium-ion batteries because it has excellent mechanical and electrical properties. Hence, the use of graphene-containing nanocomposites is suitable in LIBs.

4.2 Applications of Graphene for Particulate Removal in Aqueous Solutions

The presence of radioactive materials in water causes significant environmental and ecological problems, and various treatment methods are used to remove these materials [17].

In this context, Hu et al. [17] used powder-like graphene, carbon nanotubes, potassium permanganate, ascorbic acid, liquid uranium, and Arsenazo (III) for their studies. The experiment made it possible to prepare a hybrid hydrogel of graphene and graphene nanoribbons with a specific structure with a large surface.

In addition to the ability of the tested aerogels to the pH and the fact that the process is fast, spontaneous, and endothermic, it was discovered in this experiment.

In [18], due to the large surface area, GO was used to remove radionuclides from wastewater through their concentration and solidification. GO base naturals were used as adsorbents.

4.3 Application of Graphene as a Photodetector

It is a challenge the search for environmentally friendly synthesis routes of graphene and related materials. In this aspect, the work of Fathima et al. [19] investigates the synthesis of graphene sheets and graphene quantum dots (GQDs) from sugarcane bagasse. The obtained materials were doped with antimony and zinc oxide nanostructures to improve the optoelectrical properties of a photodetector, supported by a flexible ITO/PET substrate.

Figure 6a presents the graphene sheets' absorption spectra and the GQDs using a UV–Vis spectrophotometer. The inset shows the samples of graphene sheets and GQDs dispersed in water, under white light (left) and then under UV light (right). Strong fluorescence was exhibited by GQDs, whereas graphene sheets exhibited weak fluorescence. Figure 6b presents the Raman spectrum of graphene and GQDs, where it is possible to attain information about the defects, the nature of the bonds, and the number of layers in graphene. The results indicate that through the purposed methodology, it was possible to synthesize monolayer graphene sheets and GQDs.

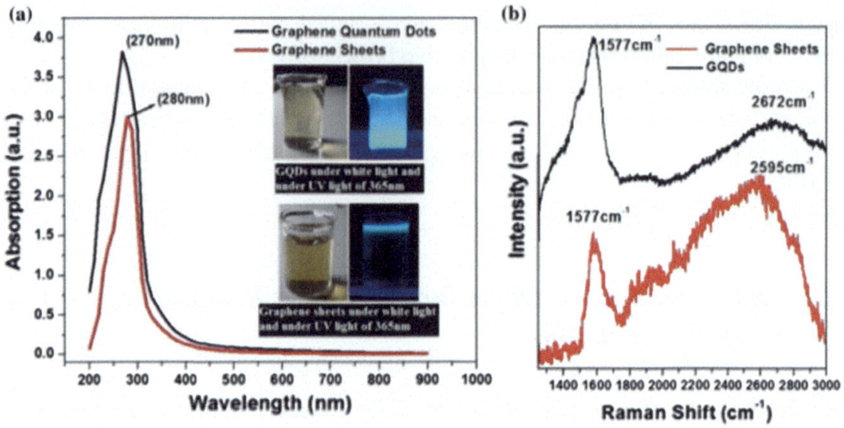

Fig. 6 **a** Absorption spectrum of graphene and graphene quantum dots (GQDs). Inset: graphene and GQDs dispersed in water under UV illumination. **b** Raman spectrum of graphene and GQDs [19]

5 Future Perspectives

The CQDs and GQDs present electronic, optical, and physical–chemical properties that offer special applications in the research of materials for optoelectronics, energy capture, chemical sensing, and biosensor [20].

For LED emitters, photodetectors, solar cells, water separation, and optoelectronic devices, the CQDs and GQDs offer in combination edge effects and confined 0D quantum structures an excellent alternative as materials [20].

Several works cited by Ghosh et al. [20] concluded that 0D structures present excellency in the optoelectronics properties, motivating several types of research on the applicability of these materials in different areas.

Examples of these promising graphene and carbon quantum dots applications are in LED, photovoltaics, photodetectors, photocatalysis, batteries, and supercapacitors.

Among the several applications points highlighted by Ghosh et al. [20] were the photodetectors using GQDs in photoconductive mode with symmetrical electrodes. Their use in solar cells was also emphasized due to their excellent optical absorption, sensitivity, and electron transport properties [20]. WLEDs for presenting a lower activation voltage using DQDs and hydrogen fuel cells are also promising uses of graphene.

References

1. Mendonça, L. H. P.: Grafeno e a sua produção a partir da grafita natural, 2018. 47f. Course conclusion work (Undergraduate degree in Mining Engineering) - Engineering School of the Federal University of Minas Gerais, Minas Gerais (2018)

2. Kar, S., Talapatra, S.: Synthesis of Graphene. Encyclopedia of Nanotechnology. Springer, Dordrecht (2016). https://doi.org/10.1007/978-94-017-9780-1_53
3. The University of Manchester, Discovery of Graphene, gra-phene.manchester.ac.uk (2014). Available in: https://www.graphene.manchester.ac.uk/learn/discovery-of-graphene/, Accessed 21 July 2021
4. Rosa, H.G.: Estudo das propriedades ópticas do grafeno e sua aplicação como absorvedor saturável em lasers à fibra dopada com érbio. 2015. 128 f. Thesis (Doctorate in electrical engineering) - Mackenzie Presbyterian University, São Paulo (2015)
5. Coroş, M., Pogăcean, F., Măgeruşan, L., et al.: A brief overview on synthesis and applications of graphene and graphene-based nanomaterials. F. Mater. Sci. **13**, 23–32 (2019). https://doi.org/10.1007/s11706-019-0452-5
6. Soldano, C., Mahmood, A., Dujardin, E.: Production, properties and potential of graphene. Carbon **48**(8), 2150 (2010). Elsevier. https://doi.org/10.1016/j.carbon.2010.01.058
7. Quintana, M., Tapia, J.I., Prato, M.: Liquid-phase exfoliated graphene: functionalization, characterization, and applications. Beilstein J. Nanotechnol. Beilstein, J. Nanotechnol. 2328–2338 (2014). https://doi.org/10.3762/bjnano.5.242
8. Nascimento, J.P.: Esfoliação Química do Grafite Natural em mistura de solventes orgânicos: a obtenção de grafenos de poucas camadas. 2013. 105f. Thesis (Master). Nuclear Technology Development Center, Minas Gerais (2013)
9. Kadari, M., Belarbi, M., Belarbi, E.H., Chaker, Y.: Nano-sensor based on ionic liquid functionalized graphene modified electrode for sensitive detection of tetrahydrofuran. In: Hatti, M. (eds.) Artificial Intelligence in Renewable Energetic Systems. ICAIRES 2017. Lecture Notes in Networks and Systems, vol 35. Springer, Cham. (2018). https://doi.org/10.1007/978-3-319-73192-6_51
10. Azizighannad, S., Mitra, S. Ss. Reduction of Graphene Oxide (GO) and its effects on chemical and colloidal properties. Sci. Rep. **8**, 10083 (2018). https://doi.org/10.1038/s41598-018-28353-6 https://creativecommons.org/licenses/by/4.0/.
11. Shaikh, T.H., Ruigang, W.: Electrochemical Exfoliation of Graphite: Effect of Temperature and Hydrogen Peroxide Addition. Electrochim. Acta **216**, 253–260 (2016). https://doi.org/10.1016/j.electacta.2016.09.022
12. Camargos, J.S.F, Semmer, A.O., Silva, S.N.: Characteristics and applications of graphene oxide and main routes for synthesis (2017). https://doi.org/10.18540/jecev13iss8pp1118-1130
13. Lu, X., Jin, X., Sun, J.: Advances of Graphene application in electrode materials for lithium ion battery. Sci. China Technol. Sci. **58**, 1829–1840 (2015). https://doi.org/10.1007/s11431-015-5927-8
14. Zhou, X., Yin, Y.X., Wan, L.J., et al.: Self-assembled nanocomposite of silicon nanoparticles encapsulated in Graphene through electrostatic attraction for lithium-ion batteries. Adv. Energy Mater. **2**, 1086–1090 (2012). https://doi.org/10.1002/aenm.201200158
15. Yi, R., Zai, J., Dai, F., et al.: Dual conductive network-enabled Graphene/Si-C composite anode with high areal capacity for lithium-ion batteries. Nano Energy **6**, 211–218 (2014). https://doi.org/10.1016/j.nanoen.2014.04.006
16. Jiang, L., Fan, Z.: Design of advanced porous graphene materials: from graphene nanomesh to 3D architectures. Nanoscale **6**, 1922–1945 (2014). https://doi.org/10.1039/C3NR04555B
17. Hu, X., Wang, Y., Wu, P., et al.: Preparation of graphene/graohene nanoribbons hybrid aerogel and its application for the removal of uranium from aqueous solutions. J. Radioanal. Nucl. Chem. **325**, 207–215 (2020). https://doi.org/10.1007/s10967-020-07208-3
18. Wang, X., Yu, S., Jin, J., et al.: Application of graphene oxides and graphene oxide-based nanomaterials in radionuclide removal from aqueous solutions. Sci. Bull. **61**, 1583–1593 (2016). https://doi.org/10/1007/s11434-016-1168-x
19. Fathima, N., Pradeep, N., Balakrishnan, J.: Green synthesis of graphene quantum dots and the dual application of graphene quantum dots-decorated flexible MSM p-type ZnO device as UV photodetector and piezotronic generator. Bull. Mater. Sci. **44** (2021). https://doi.org/10.1007/s12034-020-02326-w

20. Ghosh, D., Sarkar, K., Devi, P., Kim, K., Kumar, P.: Current and future persperspectives for carbon and graphene quantum dots: From synthesis to strategy for building optoelectronic and energy devices. Renew. Sustain. Energy Rev. **135** (2021). https://doi.org/10.1016/j.rser.2020.110391
21. Han, C., Li, H., Chen, J., Li, B., Wong, C.P.P.: Graphene-based materials with tailored nanostructures for lithium-ion batteries. In: Wong, C.P.P., Moon, K., Li, Y. (eds.) Nano-Bio-Electronic, Photonic and MEMS Packaging. Springer, Cham . 10/1007/978-3-030-49991-4_21
22. Alencar, E., Santana, D.: Processos De Obtenção Do Grafeno, Suas Aplicações E Sua Importância Para O Brasil. Oswaldo Cruz Graduate Center, São Paulo (2016). https://doi.org/10.31510/infa.v17i1.786

Tuning Nanostructured Materials Properties Through Microwave-Assisted Synthesis

Camila Stockey Erhardt

Abstract Many methods can synthesize nanomaterials. However, the methods interfere with the behavior of nanomaterials, emerging several applications. There are changes in the properties of nanomaterials through the use of microwave-assisted synthesis. The major changes are size, homogeneous distribution, and morphology modification. Changes are consequences of rapid heating provided by microwaves. These changes modify the mechanical, optical, magnetic, and electrical properties. The behavior of nanomaterials can be enhanced or depreciated depending on the parameters used in the synthesis.

Keywords Microwave-assisted synthesis · Morphology · Size · Properties

Abbreviations

GHz	GigaHertz
PANI	Polyaniline
TEM	Transmission Electron Microscopy
UV	Ultraviolet light

1 Introduction

Nanomaterials have diverse properties and applications for innovative functions. Through their size, they can have different properties than the same material of the conventional size. There are several methods of synthesis of nanomaterials [1–3]. All methods can provide a change in the properties of nanomaterials. In this way, the properties of nanomaterials are modified according to the chosen synthesis method.

C. Stockey Erhardt (✉)
Universidade Federal do Rio Grande do Sul, Porto Alegre, Brazil

A. Kopp Alves (ed.), *Technological Applications of Nanomaterials*,
Engineering Materials, https://doi.org/10.1007/978-3-030-86901-4_13

Furthermore, these properties are influenced according to the parameters determined in the synthesis.

Microwave-assisted synthesis is an example of property modification due to material structure change and an advantage in terms of speed and ease of synthesis. With these attractions and possible modifications, it is possible to enhance the properties of these nanomaterials.

Nanomaterials already have different properties from conventional materials. Changing behavior and tuning properties by synthesis give rise to new applications. These applications bring benefits to the most diverse areas, such as health, environment, and technology. The miniaturization of devices, for example, has been widely explored in the field of nanomaterials and is related to all properties.

Optical properties have varied areas of research, specially bandgap modification which thus influences photocatalysis [4] and wastewater treatment [5] research.

Finally, this chapter deals with the changes that microwave-assisted synthesis exerts on nanomaterials, altering their behavior regarding mechanical, optical, magnetic, and electrical properties.

2 Microwave-Assisted Synthesis

The microwave oven at first was intended only for heating food. Currently, this technology is used in a wide variety of research fields. The major advance that the microwave process brought was rapid heating [6].

The microwave works through a magnetron that generates a magnetic wave. Within the electromagnetic spectrum, the microwave is between 1 mm and 1 m in wavelength (0,3 a 300 GHz) [7]. Most microwaves are used at 2.5 GHz frequency [8]. In microwave-assisted synthesis, the material interacts through microwave absorption, where the material's energy generates heat [9]. This interaction can be by polarization or conduction [10]. In polarization, the main element that must exist in the matter is the dipole moment. It causes the molecule to generate heat as it aligns with the external magnetic field by rotation. This rotation creates heat. However, polarization is not the most efficient mechanism for producing heat. Conduction plays a major role in warming up. Conduction occurs when the ions in the sample due to the magnetic field move, causing collisions, and thus the kinetic energy increases, generating heat [11].

Microwave synthesis is used for rapid and homogeneous heating, where most of the time, water is used as a solvent material for the reaction [12]. As water is a bipolar material, it can absorb microwave energy and thus aid in synthesis. The most common microwave configuration can maintain a constant temperature and pressure, and the synthesis takes place inside a Teflon reactor. It is important to note that the conventional microwave oven can be used for the synthesis of materials; however, it does not maintain the necessary parameters that an industrial microwave oven has [13]. It is possible to control the heating rate until it reaches the desired temperature, power, and pressure.

One of the most studied parameters in microwave-assisted synthesis is time. It is an important parameter to understand the formation of the nanomaterial. Crystallization in microwave-assisted synthesis is much faster than in other synthesis methods. The quick reaction is the effect of superheating, which produces smaller particles and homogeneous distribution [14]. Figure 1 shows the nucleation and growth of $CuBi_2O_4$ nanoparticles that change with increasing synthesis time.

Unlike other processes, microwaves cause heating to arise from inside the material, and it is instantaneous, as is shown in Fig. 2, where conventional heating goes from the outside to the inside of the material. And it often ends up not being homogeneous. This possibility makes it different from other traditional synthesis processes that last for long periods. With the high temperature and pressure that can be achieved in a microwave-assisted synthesis, materials with innovative applications emerge. Microwave-assisted synthesis's advantages are quick heating, low cost in terms of time and energy spent, and uniformity of material [15].

Microwave-assisted synthesis brings the opportunity to modify the properties of materials, giving rise to nanostructured materials with different morphology and structures that imply changes in their applications. Therefore, the change of parameters in the synthesis such as temperature, time, and solvent changes the morphology, crystallite size, and consequently, the properties of the materials.

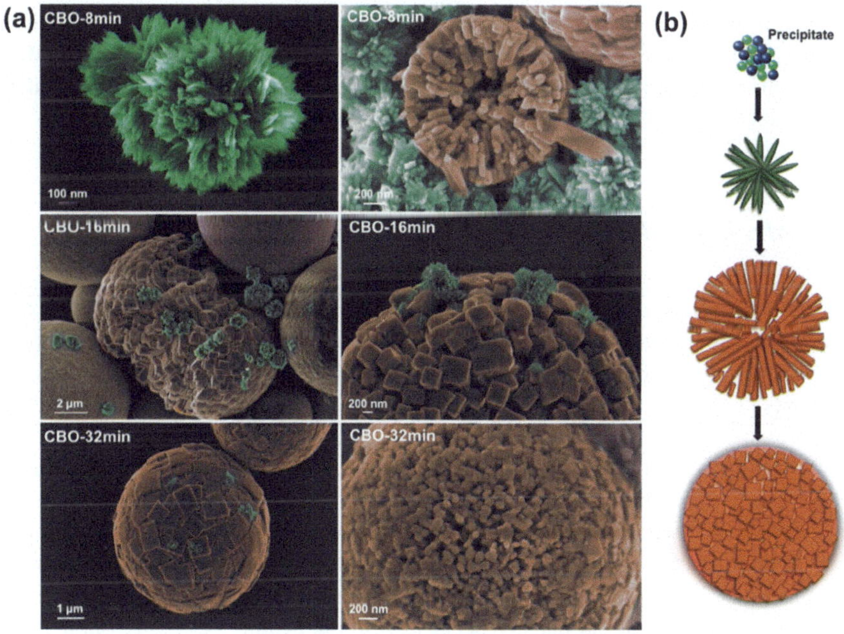

Fig. 1 FESEM micrographs of $CuBi_2O_4$ at different times of synthesis (**a**) and Schematic representation of the nucleation and growth of nanoparticles (**b**). Reprinted with permission of [16]

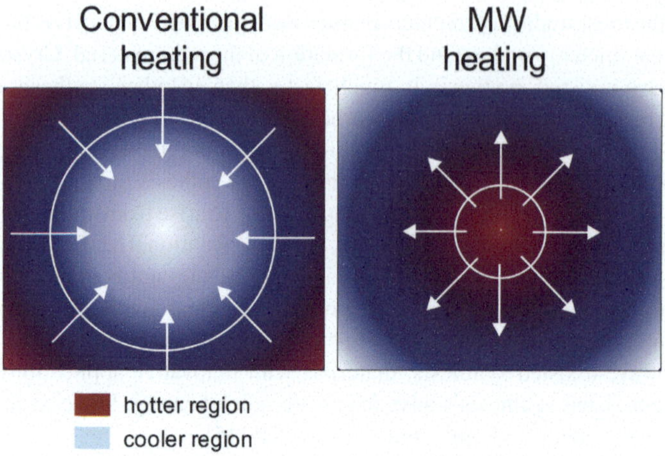

Fig. 2 Difference between conventional and microwave heating. Reprinted with permission of [11]

3 Effect of Microwave-Assisted Synthesis on Properties

This section analyzes and exemplifies what happens to the properties as a function of microwave-assisted synthesis. Applications in different areas with nanomaterials are approached. The properties dealt with here are mechanical, optical, magnetic, and electrical properties. The relationship between them and the behavior presented by nanomaterials is evident.

3.1 Mechanical Properties

In general, mechanical properties refer to the ability of materials to resist the load. Therefore, it is essential to analyze the behavior of materials for resistance depending on the application. Mechanical properties are mainly related to the microstructure of materials [17]. However, when it comes to nanomaterials, this property may differ in some aspects, not behaving in the same way as conventional materials.

With the use of microwave-assisted synthesis, there is an increase in particle size homogeneity. The increase in microstructure homogeneity and smaller particle size increases strength compared to material manufactured by another technique [18]. In addition, the microwave-assisted synthesis process helps in the reduction of nanoparticles. In this way, with the reduced grain size, fewer discordances occur. It also decreases the material density because with the heating being high in a short period, there is no increase in grain size.

Microwave-assisted synthesis is much researched in coating deposition. With the microwave method, crystals with different morphologies can grow, as is the case of a hydroxyapatite coating that formed the structure of crystals in a rod-like and flower-like morphology [19].

The resistance of nanomaterials in the biomedical area is also of great importance, as in zirconia, where structural change can lead the material to become unbalanced and fracture. Using microwave synthesis, it is possible to stabilize this nanomaterial due to a change in structure, where there are surface defects that contribute, increasing the value of toughness and hardness of biological implants [20].

3.2 *Optical Properties*

The optical properties of the material are the answer to how the material reacts to exposure to electromagnetic radiation. These properties are fundamental due to the various applications that use optical phenomena.

The use of microwaves as part of a synthesis method can modify the microstructure of the nanomaterial. This way, it affects the optical properties, improves the ceramic transparency, and can be used as an optical infrared lens [21]. One of the improvements in microwave-assisted synthesis is the modification of the bandgap in nanomaterials [22]. The reduction of the bandgap in nanomaterials brings their possibilities to function in the visible light spectrum region, making it unnecessary to use an alternative light source, such as UV light, to activate the nanomaterial in photocatalysis [23].

In some materials, as the particle size decreases, there is an increase in the bandgap. In the case of Zn_2SnO_4 nanoparticles, an increase in the surface area occurs with the size reduction. In bulk materials, the cohesive energy increases, and consequently, the bandgap increases [14]. It demonstrates that in not all cases, the bandgap decrease.

Some materials are already considered excellent photocatalysts, such as TiO_2 [24, 25] and ZnO [26, 27]. With the use of the microwave-assisted synthesis method, it is possible to improve the photocatalytic potential with the reduction of the bandgap. The synthesis also modifies the morphology and surface area of the nanomaterial [28]. Thus, making the photocatalytic process become faster, where the degradation of a contaminant that would take hours can be done in minutes [29].

Other techniques are being used through this synthesis to increase the photocatalytic potential, such as doping with other materials. Doping can be done so that the doped material modifies the microstructure of the nanomaterial, entering oxygen vacancies, or replacing other atoms in the structure. Thus, even without changing the bandgap, the material can increase the photocatalytic potential by doping [30].

3.3 *Magnetic Properties*

Magnetic materials are important, for example, because of their ability to store lots of information in a small space. Due to the need to increasingly reduce devices, nanomaterials follow this trend and store much more information than conventional material.

The magnetic properties are mainly altered by the crystallite size and its morphology in nanomaterials [31]. With the change of parameters of the method of synthesis assisted by microwave, the nanomaterial behavior can change. One of the parameters that can modify the magnetic properties is the synthesis heating time. In specific nanomaterials, with the increase of the synthesis time, the magnetic saturation may increase because, with the time increase, there is a change in morphology. The same happens with coercivity that depends on the morphology and size of crystallite [32].

Ferrites are magnetic nanomaterials and have been extensively researched. The microwave-assisted synthesis method is an alternative for the production of this nanomaterial. In general, the microwave-assisted synthesis method can be used to control the crystallite size. However, with the change of the synthesis parameters, the magnetic properties can be different, increasing or decreasing the coercivity, for example [33].

Behaviors can change for better or worse according to the synthesis parameters. For example, in $Cu_{1-x}ZnxFe_2O_4$ synthesis, magnetization decreases due to the change in cation distribution provided by microwave synthesis [34]. Also, the work of Wang et al. [35] suggests that besides microwave synthesis parameters affect the morphology of the obtained products (Fig. 3), it also affects the magnetic properties of the materials.

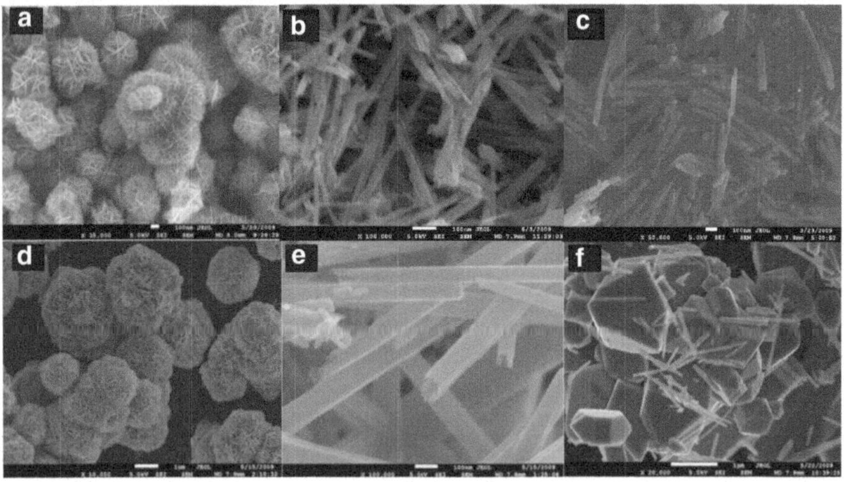

Fig. 3 FESEM images of MnO_2 nanostructures: **a** microflowers, **b** nanotubes, **c** nanorods, **d** microflowers, **e** square nanotubes, and **f** microcrystals. Reprinted with permission of [35]

3.4 Electrical Properties

The microwave-assisted synthesis technique can lower the sintering temperature and thus enhance the electrical properties. In addition, the synthesis produces highly reactive powders at low temperatures, decreasing the ion blocking effect related to grain boundaries. In this way, the ionic conductivity is higher in the nanomaterial than in the commercial material [36].

Microwave-assisted synthesis can modify the electrical behavior of the material due to the opportunity to alter its morphology through the process. For example, the consequence of the decrease in oxygen vacancies is related to morphology. Another factor that can modify the electrical behavior is the inhomogeneous distribution of crystallite sizes [37].

Nanocomposites can also be manufactured by microwave-assisted synthesis method and tuning the electrical properties. This is due to different heating than the conventional method, which reduces impurities and their change in morphology. An example of this is the MoS_2/PANI nanocomposites [38].

Microwave-assisted synthesis is an excellent synthesis method to modify the electrical property because, with the choice of parameters, it is possible to maintain the homogeneity of the crystallite size.

Solar cells are an up-and-coming application where light is converted to electrical energy relating all properties with the final behavior of the nanomaterial. As for the electrical properties, an improvement in photovoltaic performance can be seen. For example, in Fig. 4, it can be seen that after microwave-assisted synthesis, the transformation of TiO_2 spheres into nanocrystals occurs. Thus, with these spaces between nanocrystals, the surface area increases, and light-harvesting efficiency also increases [39].

4 Conclusions

Microwave-assisted synthesis has the possibility of tuning the properties of nanostructures. The main effects of the synthesis are the homogeneity of the size distribution and the modification of morphology. Therefore, it is essential to control the parameters to obtain the desired improvement of properties. The main factors of microwave-assisted synthesis are time and temperature. Knowing that all properties are related, it is possible to improve one property and worsen another.

For future research, integrating other techniques with the microwave synthesis method to enhance the effects should be analyzed. And in the future, the changes that would appear in nanomaterials transition to an industrial scale of the microwave-assisted synthesis method.

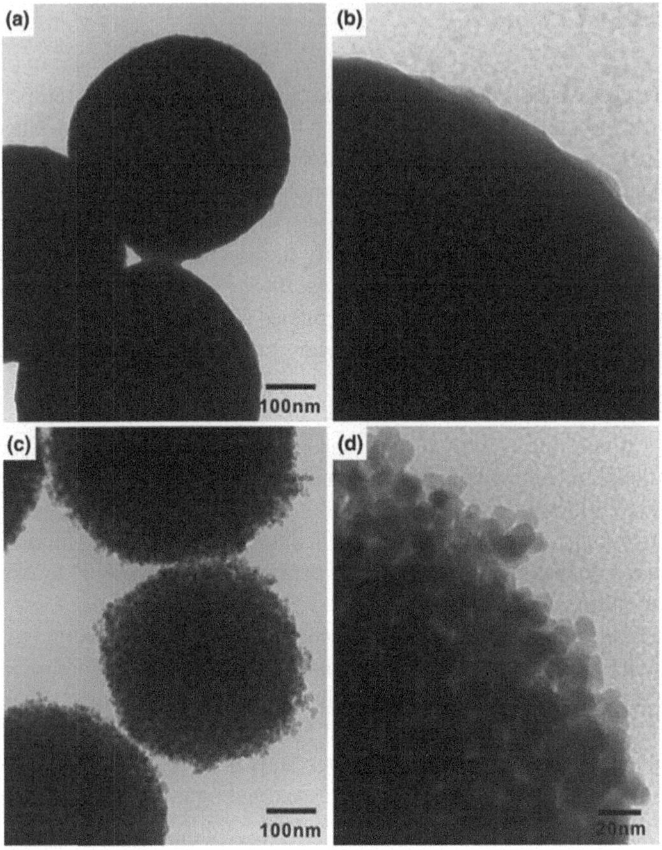

Fig. 4 TEM images of TiO_2 spheres **a** and **b** before and **c** and **d** after microwave treatment. Reprinted with permission of [39]

References

1. JJ Hinman KS Suslick 2017 Nanostructured materials synthesis using ultrasound Top. Curr. Chem. 375 1 36 https://doi.org/10.1007/s41061-016-0100-9
2. P Aarthye M Sureshkumar 2021 Green synthesis of nanomaterials: An overview Mater. Today Proc. https://doi.org/10.1016/j.matpr.2021.04.564
3. CS Erhardt LE Caldeira J Venturini SR Bragança CP Bergmann 2020 Sucrose as a sol-gel synthesis additive for tuning spinel inversion and improving the magnetic properties of $CoFe_2O_4$ nanoparticles Ceram. Int. 46 12759 12766 https://doi.org/10.1016/j.ceramint.2020.02.044
4. GP Barreto G Morales MLL Quintanilla 2013 Microwave assisted synthesis of ZnO nanoparticles: effect of precursor reagents, temperature, irradiation time, and additives on nano-ZnO Morphology development J. Mater. 2013 1 11 https://doi.org/10.1155/2013/478681
5. Liang, W., Lu, Y., Li, N., Li, H., Zhu, F.: Microwave-assisted synthesis of magnetic surface molecular imprinted polymer for adsorption and solid phase extraction of 4-nitrophenol in wastewater. Microchem. J. **159** (2020). https://doi.org/10.1016/j.microc.2020.105316

6. H Li Z Zhao C Xiouras GD Stefanidis X Li X Gao 2019 Fundamentals and applications of microwave heating to chemicals separation processes Renew. Sustain. Energy Rev. 114 109316 https://doi.org/10.1016/j.rser.2019.109316
7. LY Meng B Wang MG Ma KL Lin 2016 The progress of microwave-assisted hydrothermal method in the synthesis of functional nanomaterials Mater. Today Chem. 1–2 63 83 https://doi.org/10.1016/j.mtchem.2016.11.003
8. Lagashetty, A., Havanoor, V., Basavaraja, S.: Microwave-assisted route for synthesis of nanosized metal oxides. **8**, 484–493 (2007). https://doi.org/10.1016/j.stam.2007.07.001
9. RR Mishra AK Sharma 2016 Microwave-material interaction phenomena: Heating mechanisms, challenges and opportunities in material processing Compos. Part A Appl. Sci. Manuf. 81 78 97 https://doi.org/10.1016/j.compositesa.2015.10.035
10. Wang, S.: Microwave processing. Handb. Food Saf. Eng. 371–393 (2012). https://doi.org/10.1002/9781444355321.ch15
11. PP Falciglia P Roccaro L Bonanno G Guidi De FGA Vagliasindi S Romano 2018 A review on the microwave heating as a sustainable technique for environmental remediation/detoxification applications Renew. Sustain. Energy Rev. 95 147 170 https://doi.org/10.1016/j.rser.2018.07.031
12. Y Sun P Zhang J Hu B Liu J Yang S Liang K Xiao H Hou 2021 A review on microwave irradiation to the properties of geopolymers: Mechanisms and challenges Constr. Build. Mater. 294 123491 https://doi.org/10.1016/j.conbuildmat.2021.123491
13. JA Nóbrega LC Trevizan GCL Araújo ARA Nogueira 2002 Focused-microwave-assisted strategies for sample preparation, Spectrochim Acta - Part B At. Spectrosc. 57 1855 1876 https://doi.org/10.1016/S0584-8547(02)00172-6
14. Reyes, O., Pal, M., Escorcia-García, J., Sánchez-Albores, R., Sebastian, P.J.: Microwave-assisted chemical synthesis of Zn_2SnO_4 nanoparticles. Mater. Sci. Semicond. Process. **108** (2020). https://doi.org/10.1016/j.mssp.2019.104878
15. A Kumar Y Kuang Z Liang X Sun 2020 Microwave chemistry, recent advancements, and eco-friendly microwave-assisted synthesis of nanoarchitectures and their applications: a review Mater. Today Nano. 11 100076 https://doi.org/10.1016/j.mtnano.2020.100076
16. LS Ribeiro IM Pinatti JA Torres AS Giroto F Lesse E Longo C Ribeiro AE Nogueira 2020 Rapid microwave-assisted hydrothermal synthesis of $CuBi_2O_4$ and its application for the artificial photosynthesis Mater. Lett. 275 128165 https://doi.org/10.1016/j.matlet.2020.128165
17. Sathiyamoorthi, P., Kim, H.S.: High-entropy alloys with heterogeneous microstructure: Processing and mechanical properties. Prog. Mater. Sci. 100709 (2020). https://doi.org/10.1016/j.pmatsci.2020.100709
18. Ghomash Pasand, E., Nemati, A., Solati-Hashjin, M., Arzani, K., Farzadi, A.: Microwave assisted synthesis & properties of nano HA-TCP biphasic calcium phosphate. Int. J. Miner. Metall. Mater. **19**, 441–445 (2012). https://doi.org/10.1007/s12613-012-0576-4
19. S Shen S Cai M Zhang G Xu Y Li R Ling X Wu 2015 Microwave assisted deposition of hydroxyapatite coating on a magnesium alloy with enhanced corrosion resistance Mater. Lett. 159 146 149 https://doi.org/10.1016/j.matlet.2015.06.096
20. T Batool BS Bukhari S Riaz KM Batoo EH Raslan M Hadi S Naseem 2020 Microwave assisted sol-gel synthesis of bioactive zirconia nanoparticles—correlation of strength and structure J. Mech. Behav. Biomed. Mater. 112 104012 https://doi.org/10.1016/j.jmbbm.2020.104012
21. Z Tang Z Huang W Han J Qi N Ma Y Zhang T Lu 2020 Microwave-assisted synthesis of uranium doped $Y_2Zr_2O_7$ transparent ceramics as potential near-infrared optical lens Scr. Mater. 178 90 93 https://doi.org/10.1016/j.scriptamat.2019.11.007
22. TL Ruwer J Venturini S Khan CP Bergmann 2020 Quick synthesis of homogeneous Nb_2O_5 nanorod arrays via a microwave-assisted hydrothermal method Mater. Lett. 265 127429 https://doi.org/10.1016/j.matlet.2020.127429
23. M Parthibavarman S Sathishkumar M Jayashree R BoopathiRaja 2019 Microwave assisted synthesis of pure and Ag doped SnO_2 quantum dots as novel platform for high photocatalytic activity performance J. Clust. Sci. 30 351 363 https://doi.org/10.1007/s10876-018-01493-5

24. AG Agrios P Pichat 2005 State of the art and perspectives on materials and applications of photocatalysis over TiO_2 J. Appl. Electrochem. 35 655 663 https://doi.org/10.1007/s10800-005-1627-6
25. Al-Mamun, M.R., Kader, S., Islam, M.S., Khan, M.Z.H.: Photocatalytic activity improvement and application of UV-TiO_2 photocatalysis in textile wastewater treatment: A review. J. Environ. Chem. Eng. **7** (2019). https://doi.org/10.1016/j.jece.2019.103248
26. X Chen Z Wu D Liu Z Gao 2017 Preparation of ZnO photocatalyst for the efficient and rapid photocatalytic degradation of Azo dyes Nanoscale Res. Lett. 12 4 13 https://doi.org/10.1186/s11671-017-1904-4
27. K Qi B Cheng J Yu W Ho 2017 Review on the improvement of the photocatalytic and antibacterial activities of ZnO J. Alloys Compd. 727 792 820 https://doi.org/10.1016/j.jallcom.2017.08.142
28. Akshatha, S., Sreenivasa, S., Parashuram, L., Udaya kumar, V., Alharthi, F.A., Chakrapani Rao, T.M., kumar, S.: Microwave assisted green synthesis of p-type Co_3O_4@Mesoporous carbon spheres for simultaneous degradation of dyes and photocatalytic hydrogen evolution reaction. Mater. Sci. Semicond. Process **121**, 105432 (2021). https://doi.org/10.1016/j.mssp.2020.105432
29. S Choudhary A Bisht S Mohapatra 2021 Microwave-assisted synthesis of α-Fe_2O_3/$ZnFe_2O_4$/ZnO ternary hybrid nanostructures for photocatalytic applications Ceram. Int. 47 3833 3841 https://doi.org/10.1016/j.ceramint.2020.09.243
30. CH Claudino M Kuznetsova BS Rodrigues C Chen Z Wang M Sardela JS Souza 2020 Facile one-pot microwave-assisted synthesis of tungsten-doped $BiVO_4$/WO_3 heterojunctions with enhanced photocatalytic activity Mater. Res. Bull. 125 110783 https://doi.org/10.1016/j.materresbull.2020.110783
31. UT Nakate SN Kale 2016 Microwave assisted synthesis and characterizations of $NiCo_2O_4$ nanoplates and Electrical, magnetic properties Mater. Today Proc. 3 1992 1998 https://doi.org/10.1016/j.matpr.2016.04.101
32. Z Ai K Deng Q Wan L Zhang S Lee 2010 Facile microwave-assisted synthesis and magnetic and gas sensing properties of Fe_3O_4 nanoroses J. Phys. Chem. C. 114 6237 6242 https://doi.org/10.1021/jp910514f
33. Gurgel, A.L., Martinelli, A.E., de A. Conceição, O.L., Xavier, M.M., Morales Torres, M.A., de Araújo Melo, D.M.: Microwave-assisted hydrothermal synthesis and magnetic properties of nanostructured cobalt ferrite. J. Alloys Compd. **799**, 36–42 (2019). https://doi.org/10.1016/j.jallcom.2019.05.309
34. Tony Dhiwahar, A., Sundararajan, M., Sakthivel, P., Dash, C.S., Yuvaraj, S.: Microwave-assisted combustion synthesis of pure and zinc-doped copper ferrite nanoparticles: Structural, morphological, optical, vibrational, and magnetic behavior. J. Phys. Chem. Solids. **138**, 109257 (2020). https://doi.org/10.1016/j.jpcs.2019.109257.
35. J Wang R Zeng G Du W Li Z Chen S Li Z Guo S Dou 2015 Rapid microwave-assisted synthesis of various MnO_2 nanostructures and their magnetic properties Mater. Chem. Phys. 166 42 48 https://doi.org/10.1016/j.matchemphys.2015.08.020
36. JDC Carregosa JPF Grilo GS Godoi DA Macedo RM Nascimento RMPB Oliveira 2020 Microwave-assisted hydrothermal synthesis of ceria (CeO_2): Microstructure, sinterability and electrical properties Ceram. Int. 46 23271 23275 https://doi.org/10.1016/j.ceramint.2020.06.021
37. Ferrel-Alvarez, A.C., Domínguez-Crespo, M.A., Cong, H., Torres-Huerta, A.M., Palma-Ramírez, D., Irvine, J.T.S.: Microwave irradiation synthesis to obtain La0.7-xPrxCa$_{0.3}$$MnO_3$ perovskites: Electrical and electrochemical performance. J. Alloys Compd. **851**, 156882 (2021). https://doi.org/10.1016/j.jallcom.2020.156882.
38. S Krithika J Balavijayalakshmi 2021 Synthesis and fabrication of nanostructured MoS_2/PANI nanocomposites by microwave assisted method for electrochemical applications Mater. Today Proc. https://doi.org/10.1016/j.matpr.2021.01.013
39. Chen, P., De Peng, J., Liao, C.H., Shen, P.S., Kuo, P.L.: Microwave-assisted hydrothermal synthesis of TiO_2 spheres with efficient photovoltaic performance for dye-sensitized solar cells. J. Nanoparticle Res. **15** (2013). https://doi.org/10.1007/s11051-013-1465-0.

The Role of Glycerol in the Synthesis of Nanomaterials

João Vitor Braun and Annelise Kopp Alves

Abstract Among the newest challenges in nanomaterials, development is the search for new methods of synthesis to produce complex structures in a controllable way. Furthermore, finding new green raw materials and reactants is also something necessary to support sustainable development goals. Glycerol is a biobased product, largely available, biodegradable, and presents properties favorable to produce structured nanoparticles. This chapter will discuss the role played by glycerol as a precursor to obtaining different kinds of nanoparticles.

Keywords Glycerol · Glycerolate · Hollow nanostructure · Hierarchical structure

Abbreviations

MW	Microwave
CaGL	Calcium glycerolate
CoGL	Cobalt glycerolate
FeGL	Iron glycerolate
InGL	Indium glycerolate
LiGL	Litium glycerolate
MnGL	Manganese glycerolate
NiGL	Nickel glycerolate
SEM	Scanning Electron Microscopy
SEM-EDS	Scanning Electron Microscopy—Energy Dispersive Spectroscopy
TEM	Transmission Mlectron Microscopy
TiGL	Titanium glycerolate
VGL	Vanadium glycerolate
XPS	X-ray photoelectron spectroscopy
ZnGL	Zinc glycerolate
ZrGL	Zirconium glycerolate

J. V. Braun (✉) · A. K. Alves
Universidade Federal Do Rio Grande Do Sul, Porto Alegre, Brazil

A. Kopp Alves (ed.), *Technological Applications of Nanomaterials*,
Engineering Materials, https://doi.org/10.1007/978-3-030-86901-4_14

1 Introduction

When the C60 molecule was identified in 80s the studies about nanostructures were still not mainstream inside the scientific community. At that time, mass spectrometry was starting to emerge as a revolutionary instrumental method, and it took some time until the classic football shape of fullerene was unveiled. However, shortly thereafter, in 1990, the German chemist Wolfgang Krätschmer, the US chemist Donald Huffman, and their students had success isolating C60 molecules from carbon soot in bulk, thereby making this substance available for large-scale experiments. That progress brought the study of carbon nanostructures to the target of researchers, who in a short time started to use them as starter blocks to produce innovative materials such as carbon nanotubes and graphene [1].

Since then, uncountable breakthroughs in instrumental techniques and synthesis methods allowed nanostructures to emerge with the most creative and unconventional shapes [2, 3]. From that variety of morphologies, a series of uncommon and unexpected properties arose, leading to the challenge of producing homogeneous and pure particles, controlling the shape, and dragging out the highest performance possible.

Figure out methods to enable the production of high-quality materials through new green solvents is mandatory to build a sustainable pathway to the future [4, 5]. The solvent plays a key role in the quality of the final product, and glycerol has demonstrated success as raw material in the synthesis of high-purity nanoparticles.

The glycerol is a coproduct from the biodiesel industry, which places it as a potential source to supply the green chemistry market [6]. Besides that, glycerol is non-toxic, non-inflammable, and biodegradable, also presents a high solubility, low vapor pressure, and high boiling point. Finally, glycerol can be employed as a soft template or as a chelation agent to synthesize metal oxides.

In the form of a metallic salt (glycerolate), it possesses the ability to create stable crystalline structures able to retain the original form of the particle even after removing the organic part of the material [7].

Along this text, we will wander a few examples of how glycerol was already successfully applied in the synthesis of nanomaterials and how the glycerolate salts present intriguing properties by themselves.

2 Metal Glycerolate as Precursors of Inorganic Nanoparticles

Titanium Glycerolate

One important material largely studied in the last years is TiO_2. With the ability to perform in different fields due to unidirectional electron transport, nanoparticles of TiO_2, such as nanowires and nanorods, have demonstrated outstanding potential to be applied in different areas, from water treatment to energy production. There are several methods to obtain those forms of TiO_2, but there are not many studies about glycerol-based reaction mediums.

In 2015, Bae and Jang [8] demonstrated how the TiO_2 anatase crystals could be obtained from Titanium glycerolate (TiGL), keeping almost the exact morphology of the glycerol-based precursor.

Using a precipitation process under microwave (MW) heating, they showed that glycerol could direct the structure of the TiGL crystals from small rod-like shapes—around 0,08 nm—to hedgehog-shapes bigger than 0,2 nm. Furthermore, high-purity TiO_2 anatase crystals were obtained after an MW-assisted hydrolysis process that removed all the organic parts of the particle. This method makes it possible to produce TiO_2 nanoparticles keeping almost the same morphology of the glycerol-based precursor.

Another determinant point clarified by the authors is that the final shape of the particles is determined by the fast or slow growth of TiGL crystals. The molar ratio of glycerol can be used to control it, and using a cosolvent is possible to change the glycerol concentration in the reaction media. For example, when the authors reduced the glycerol concentration using isopropanol, the Titanium glycerolate crystallization took place slowly, favoring the growth of a hierarchical structure with a hedgehog form. It is also important to note that no ionic additives or surfactants were needed to orientate the shapes and that the glycerol can form a stable structure by itself.

The MW-assisted synthesis has been showing excellent results in nanomaterials production. For example, Vinodhini et al. [9] synthesized TiO_2 rod-like structures and nanoparticles using a domestic microwave oven under different microwave powers, showing how this glycerol medium is suitable for that condition without any additives.

When the MW power was increased, a considerable change in the morphology was noticed. As shown in Fig. 1, at low power (240 W), spheres of Titanium glycerolate were formed. Those spheres started to stack up when the power was increased to 480 W, forcing a 1D growth of a rod-like structure with a diameter of 33 nm. Continuing up to 720 W, spherical nanoparticles started to nucleate on the nanorod's surface, creating a double composed structure. Finally, at 960 W, the form of the particles changed entirely to a 3-D flower-like structure.

Unlike what was demonstrated by Bae and Jong, where the organic part is removed through a hydrolysis process, the annealing process cannot always maintain the shape built by the primary structure. For example, it was demonstrated by Vinodhini et al. that after annealing the samples at 450 °C, only the rod-like form is retained, and even so, presenting a considerable variation on the diameters of the rods—from 30 to 104 nm.

Since glycerol can compose a wide range of solvent mixes, that parameter can be used to create a reaction media intended to produce fine and complex nanostructures. For example, in the procedure described by Li et al. in 2013 [10], ethanol, ether, and glycerol were stirred with $TiOSO_4.2H_2O$ for 36 h, transferred into a 100 ml autoclave kept at 140 °C for 48 h. After washing, drying, and calcination at 600 °C, a high purity anatase tubular structure was obtained. The hierarchical structure built by nanolayers formed nanotubes with an inner diameter of around 500 nm.

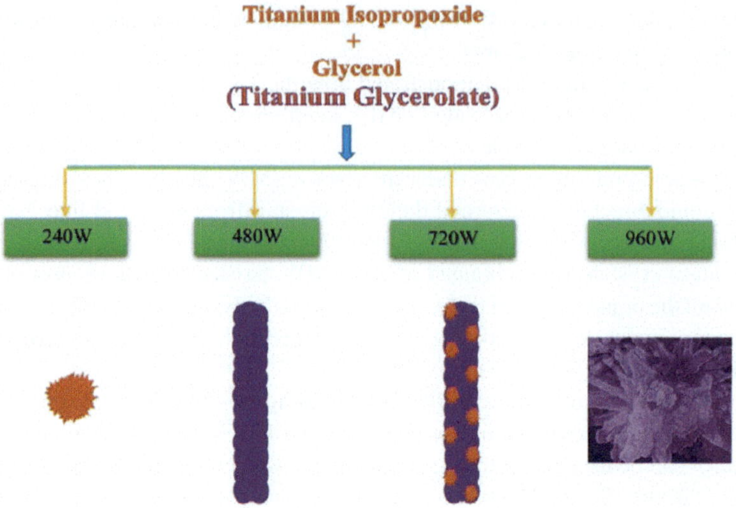

Fig. 1 Schematic illustration of morphology evolution summary as a function of microwave power for titanium glycerolate. Edited and reprinted with permission from [9]

Through SEM and TEM images analyzed in different stages of the synthesis, it is possible to understand the growth mechanism of nanotubes. The TiO_2SO_4 precursor has a smooth and solid fiber shape with a diameter from 0,3 μm to 2 μm (Fig. 2a). After 3 h of reaction, numerous small sheets arise on the surface of the fibers, then the surface becomes coarse but still in a solid shape. A hollow urchin-like structure takes the place of the fibers after 12 h of solvothermal reaction (Fig. 2c). Finally, after 48 h of reaction, the nanosheets grow larger (Fig. 2d).

Experimental attempts using only ethanol as a solvent produced Titanium nanospheres, proving the crucial role of glycerol in the obtention of the desired form. Furthermore, to reinforce the importance of the particle shape in the final properties, the unique mesoporous hierarchical tubular structure showed enhanced photocatalytic activity superior to Degussa P25, besides a high specific capacity and excellent cycling stability when used as an anode material for lithium-ion batteries.

Manganese and Cobalt Glycerolates

Metal glycerolates are the ideal precursors not only to control the desired final shape but also to produce mixed oxides with variable metallic rates. To achieve that, the temperature control of thermal treatment is a key factor because, at the same time that the temperature needs to be high enough to decompose the organic part of the molecule; extreme temperatures can affect the crystallinity and final shape of the material.

Li et al. [11] demonstrated how metal defects affect the electronic structure, improving the oxygen reduction reaction efficiency. For example, glycerol can create a reaction medium to introduce Mn and Co vacancies *in-situ* into spinel Mn_xCo_3-XO_4 through a simple solvothermal synthesis followed by thermal decomposition at

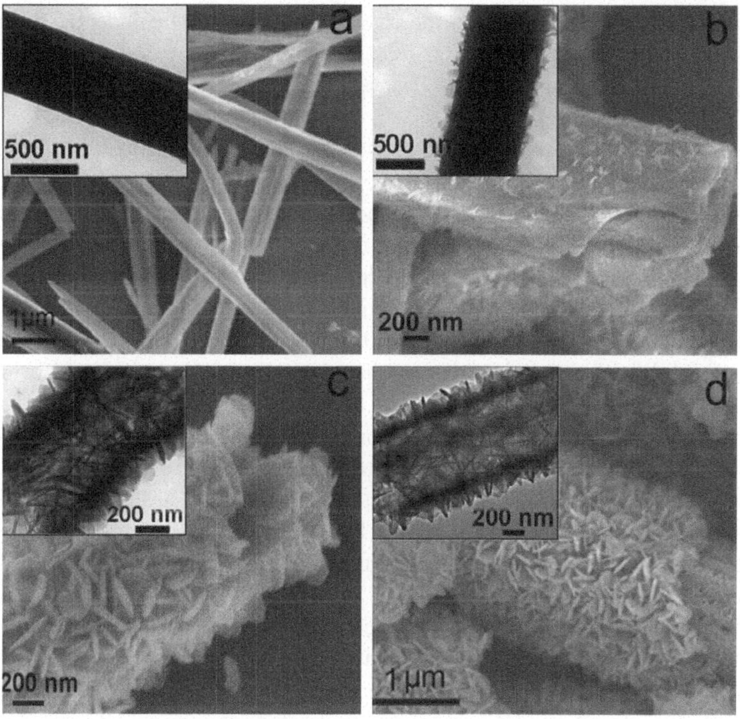

Fig. 2 SEM and TEM images of starting material $TiOSO_4H_2O$ (**a**) and samples solvothermally reacted for 3 h (**b**), 12 h (**c**), and 48 h (**d**). All samples were prepared without calcination. Edited and reprinted with permission from [10]

300 °C under air atmosphere. In mixed oxides like that, different surface and bulk compositions were formed due to the kinetics of precipitation and crystal growth of Manganese-Cobalt glycerolates. In that case, an Mn-rich surface is formed since the Manganese glycerolate (MnGL) production increases as the reaction proceeds, leading to a surface metal rate of 1:1, while in bulk, this rate is Mn:Co equal to 1:2 ($Mn_{1.5}Co_{1.5}O_4$ and $MnCo_2O_4$).

Jankovsky et al. [12] reported the production of Mn oxide nanocrystal clusters by thermal decomposition of MnGL. Pure MnO nanocrystalline was produced when MnGL is thermally decomposed under an inert atmosphere. However, when the annealing process occurs exposed to air, different oxidation states of Mn arise depending on the temperature.

At 477 °C, a single phase of Mn_3O_4 was produced. Then, a Mn_2O_3 phase appears when the temperature increases until 577 °C, forming a mix of these two crystalline phases in the nanoparticles. Finally, when the temperature reaches 677 °C, only Mn_2O_3 remains as a single phase. Although the Mn_2O_3 phase should be thermodynamically more stable than Mn_3O_4 at low temperatures, the presence of glycerolate ion and its oxidation process affects the kinetics oxidation of Mn hinder the formation of the most stable nanoparticle.

Vanadium Glycerolate

Like nanotubes, hollow spheres have unusual and interesting properties that can be used to achieve new levels of materials performance. For example, Guo et al. [13] demonstrated that through a two-step process: solvothermal synthesis of Vanadium glycerolate (VGL), followed by thermal transformation, double or single-shelled hollow spheres of V_2O_5 can be obtained. This is possible by controlling the annealing temperature at 350 °C and 450 °C, respectively.

The Vanadium glycerolate consists of 500 nm solid spheres with a smooth surface. Due to the ability of glycerolate precursors to be entirely eliminated at low temperatures, a V_2O_5 well-crystallized phase can be produced, retaining the same shape as the VGL precursor. After the thermal decomposition, the smooth surface of VGL is turned into a rough surface composed of interconnected V_2O_5 nanocrystals.

The shells are composed of nanoplates with a diameter around 100 nm and a thickness between 20 and 30 nm. In the Fig. 3a, it is possible to see from some broken spheres the core–shell structure formed. The TEM image (Fig. 3b) testify the gaps between the outer and inner shells.

Not only the temperature but also the exposure time to the thermal treatment is essential to achieve the desired final form of the particles. The oxidation of VGL begins from the surface to the center, creating a rougher surface and turning to a shell while the core continues as a solid particle along the time. Only after 60 min, unique double-shelled hollow spheres were successfully obtained at 350 °C.

Since the oxidation of VGL starts at the surface of the particle, a possible mechanism for the formation of the hollow spheres is the actuation of two oppositely directed forces: a contraction force induced by the decomposition of the organic part of glycerolate and the adhesion force induced by the simultaneous formation of V_2O_5 crystallite agglomerates which prevent inward contraction.

Fig. 3 SEM images of **a** fresh ZnO nanostars before the first use, and **b** ZnO nanostars after five catalytic cycles presenting the same star-like morphology but with changes on the surface. Edited and reprinted with permission from [16]

Iron Glycerolate

The particle size control is essential for tailoring the nanomaterial magnetic properties, for example. We already know that the crystallite growth and the final shape of the particles can be controlled by the temperature of glycerolate decomposition, in addition to the atmosphere involved in the process.

A simple method to obtain metal glycerolates is boiling glycerol with a metal source under reflux for a few hours. Usually, this method produces amorphous particles, but through those amorphous glycerolate precursors, it is possible to obtain high-quality metal oxide nanoparticles after the correct thermal treatment.

Barturek et al. [14] demonstrated this effect for CoFe mixed oxides obtained from large agglomerates of Cobalt-Iron glycerolates (CoFeGL). After the synthesis, the CoFeGL particles presented sizes of about 20 m. After the thermal treatment at 500 and 600 °C, the glycerol-based precursor was converted to ultrafine nanoparticles with a homogeneous size distribution between 3 and 12 nm and 6–20 nm, respectively.

When the temperature increases to 700 °C, the range rises to 20–40 nm, and 35–95 nm at 800 °C. The nanosizing effect is noticed in the magnetics experiments, where the fine size distribution leads to better magnetics properties.

It is also important to notice that through this simple method, boiling the reactants under reflux; it is possible to control the stoichiometry and the chemical composition of the final product. Raman, SEM–EDS, and XPS can prove the absence of any contaminant, such as Na^+ or other ions, and residual surfactant, usually left by hydrothermal and co-precipitation synthesis.

Indium Glycerolate

The employment of glycerolate as a template-free route to synthesize nanoparticles is relatively new. One of the interesting structured configurations that can be obtained from glycerolates precursors is ultrathin materials. These kinds of materials have a wide potential of application and have received attention since the graphene discovery. For example, 3D structures formed by overlapped nanosheets of InGL were synthesized by Cao et al. [15] in a mixture of glycerol, isopropanol, and $InNO_3$.

Following the evolution of the InGL structure along the time, it is possible to notice the phenomenon of self-template, where the InGL acts both as a template and precursor to the final nanosheets 3D structure. At the first hour of reaction, the material is a solid sphere with 380 nm of diameter. At the second hour of reaction, many thin nanosheets start to arise on the surface of the particle. The nanosheets growth continues along the time, making the hollow sphere bigger. Finally, after 6 h of reaction, a well-defined 3D structure formed by nanosheets is formed, measuring around 880 nm. Figure 4 presents a TEM image sequence showing how the structure evolves from solid spheres to spheres formed by nanosheets of 2 nm of thickness.

Zinc Glycerolate

ZnO is an exceptional heterogeneous catalyst for biodiesel production. It is extremely robust and can act simultaneously in esterification and transesterification reactions, especially for its unique nanostar form (Fig. 3). In the transesterification reaction of

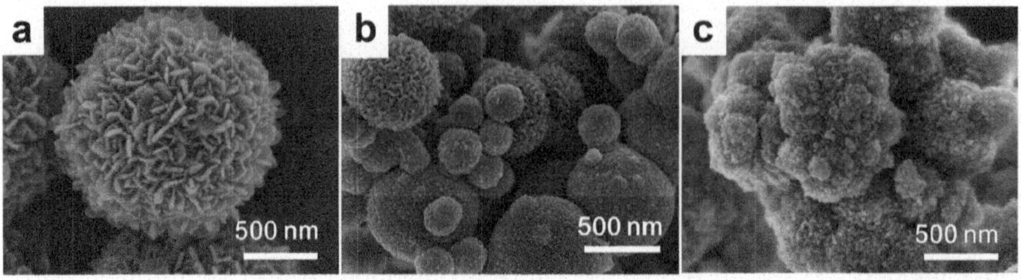

Fig. 4 SEM images of **a** FeGL, **b** $Fe_{0.5}Ni_{0.5}GL$, and **c** NiGL. Edited and reprinted with permission from [18]

canola oil with methanol, the performance of the ZnO catalyst increases significantly from 67,7% on the first use to 90,0% after five cycles. For the same reaction, ZnGL presents a conversion of 94,6%, which suggests that the improvement is caused by the formation of ZnGL in-situ, which deposits along the surface acting as a catalyst disperse over a star-like ZnO support. After the fifth use (Fig. 4b), the SEM image of ZnO shows the catalyst surface is slightly corroded but maintains its original star-like nanostructure [16].

3 Direct Uses of Glycerolate Nanoparticles

Medical Applications

Although metal glycerolate nanoparticles present such interesting properties by themselves, their application as a final product is still limited. A field that has attracted interest in applying nanotechnology is the cosmetic and dermatology industry. But as with every new discovery in the health field, a large evaluation of the potential health hazards, toxicity, and effects in the long term is fundamental. The control of morphology is an essential step to achieve this knowledge, producing materials with stable and predictable properties.

With the ability to act as a slow-release non-toxic source of therapeutic Zinc, Zinc glycerolate (ZnGL) is also a promissory antiarthritic, antiulcer, and anti-inflammatory agent. Remias et al. [17] showed an interesting method to obtain ZnGL nanorods starting from ZnO nanorods as sacrificial templates. This approach is really interesting because, ideally, the morphology of the reactants should be mirrored in the structure of the product. Still, it is a condition hard to achieve, especially when the reaction involves the complete chemical restructuring of both reactants.

In that study, the authors compare ZnGL when obtained from commercial amorphous particles of ZnO, and from ZnO nanorods synthesized by the authors. The difference is clear: while the random heap of hexagonal prisms is obtained without a well-defined form ZnO template, hexagonal prisms of glycerolate were formed aligning face-to-face to create microstacks.

One crucial point to notice is that, in this case, the reaction occurs in a static reaction medium. This means that the hydrothermal growth of micro stacks takes place in a diffusion-limited environment, which favors the attack by the cheating agent in the tips of the nanorods. Furthermore, this condition leads to the spontaneous formation of hexagonal prisms, bringing up another crucial parameter to the nanoparticles shape control.

Catalysis

As already mentioned before, the stoichiometric control between two different metals is easily controllable. In mixed metal glycerolates, the content of each metal is responsible for the final shape and structure of the particles. This phenomenon directly affects the properties of those materials; for example, Iron-Nickel glycerolate (FeNiGL) presents high performance when applied as a catalyst for Oxygen Evolution Reaction compared with the singular metal glycerolates [18].

The changes in the particle structure related to the amount of each metal are shown in Fig. 4. The pure FeGL sample comprises well-defined uniform flower-like spheres built of numerous two-dimensional nanosheets (Fig. 4a). On the opposite side (Fig. 4c), NiGL morphology shows a solid coarse surface and looks like a nanoparticle-aggregated coral microstructure. The high-angle annular dark-field scanning TEM confirms the solid nature of spheres, indicating the complete absence of hierarchical architecture. Naturally, the mixed glycerolate structure is something between the respective single glycerolates (Fig. 4b).

Another metal glycerolate with excellent catalytic performance is Calcium glycerolate (CaGL). [19–21] Kouzu et al. [22] demonstrated that the formation *in-situ* of CaGL during the transesterification reaction catalyzed by CaO had enormous influence over reaction kinetics. That potential of glycerolate ion arouses interest in other metal glycerolates, such as ZnGL [23, 24], LiGL [25], FeGL, CoGL [26], and MnGL [27], as heterogeneous catalysts for biodiesel production.

A big challenge faced for these materials when applied as catalysts is post-reaction stability. When comparing SEM images of CaGL before and after being used as a catalyst to the transesterification reaction of Soybean oil with methanol [28], considerable changes in its structure are clearly noticed. In the fresh form, CaGL is composed of clean agglomerates of prims (Fig. 5a). However, after 90 min of reaction, the agglomerates became covered with a layer of reactants and side products, mainly glycerol. As a result, the morphology of the particles changes to a non-defined and amorphous surface (Fig. 5b) with variable sizes and shapes.

4 Final Considerations

TiO_2/Carbon composites are a creative way to overcome the limitations founded by TiO_2 in its wide range of applications. There are several ways to synthesize that kind of material, in special methods where the Titanium and Carbon sources come from the used precursors. For example, Zakharova et al. [29] recently presented a

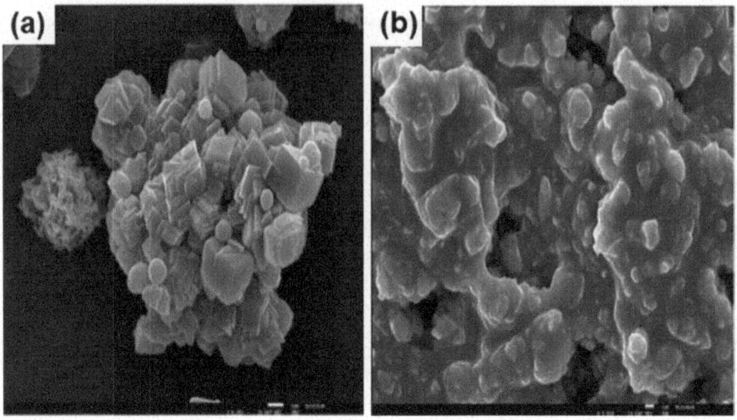

Fig. 5 SEM images of **a** Fresh CaGL, and **b** CaGL after 90 min reaction. Edited and reprinted with permission from [28]

study about morphology, structure, and phase transformation of a TiO_2/C composite obtained through the pyrolysis of TiGL.

Also, Magnetic Solid Phase Extraction Technology has been gained space on the separation of biomarker molecules. This process avoids a slow centrifugation step; however, the low chemically activity of Fe_3O_4 1D porous nanostructures makes hard the immobilization of some molecules. Aiming to circumvent this limitation, Xu et al. [30] suggested using Fe3O4 magnetic nanoparticles encapsulated with ZrGL nanowires to extract urinary ribonucleosides [30].

In that work, the authors synthesized those two different molecules (Fe_3O_4 and ZrGL) separately and summited both to a hydrothermal synthesis, creating the product Fe_3O_4@ZrGL. Although the adsorption capacity was not exceptional in this specific case, a versatile approach to immobilize metal glycerolates in magnetic particles was demonstrated for applications in metabolomics research.

Therefore, it is noticeable the great potential of glycerol-based reactions to expand the nanomaterials synthesis in the next few years. Therefore, promoted for the extensive availability and friendly properties, it is expected that in the future, glycerol will have a key role as a raw material for the synthesis of nanoparticles and as a base to discover new catalytic and functional materials.

References

1. Ajayan, P.M.: The nano-revolution spawned by carbon. Nature **575**, 49–50 (2019). https://doi.org/10.1038/d41586-019-02838-4
2. Guo, T., et al.: A comprehensive review on synthesis methods for transition-metal oxide nanostructures. Cryst. Eng. Commun. **17**, 3551–3585 (2015). https://doi.org/10.1039/c5ce00034c

3. Chen, Y., et al.: Two-Dimensional Metal Nanomaterials: Synthesis, Properties, and Applications. Chem. Rev. **118**, 6409–6455 (2018). https://doi.org/10.1021/acs.chemrev.7b00727
4. Checa, M., et al.: Recent advances in glycerol catalytic valorization: A review. Catalysts **10**, 1279 (2015). https://doi.org/10.3390/catal10111279
5. García, J.I. et al.: Green solvents from glycerol. Synthesis and physico-chemical properties of alkyl glycerol ethers. Green Chem. **12**, 426–434 (2010). https://doi.org/10.1039/b923631g
6. Pradhan, S., et al.: Synthesis of potassium glyceroxide catalyst for sustainable green fuel (biodiesel) production. J. Ind. Eng. Chem. **46**, 266–272 (2017). https://doi.org/10.1016/j.jiec.2016.10.038
7. Zhao, J., et al.: Self-template construction of hollow Co3O4 microspheres from porous ultrathin nanosheets and efficient noble metal-free water oxidation catalysts. Nanoscale **6**, 7255–7262 (2014). https://doi.org/10.1039/c4nr00002a
8. Kim, H.-B., Jang, D.-J.: Cryst. Eng. Commun. **17**, 3325–3332 (2015). https://doi.org/10.1039/c5ce00257e
9. Vinodhini, J., et al.: Effect of microwave power irradiation on TiO_2 nanostructures and binder free paste screen printed dye sensitized solar cells. Cer. Int. **45**, 4667–4673 (2019). https://doi.org/10.1016/j.ceramint.2018.11.157
10. Li, Y., et al.: Hierarchical tubular structure constructed by mesoporous TiO_2 nanosheets: Controlled synthesis and applications in photocatalysis and lithium ion batteries. Chem. Eng. J. **232**, 356–363 (2013). https://doi.org/10.1016/j.cej.2013.07.124
11. Li, K., et al.: Metal-defected spinel MnxCo$_{3\text{-}x}$O$_4$ with octahedral Mn-enriched surface for highly efficient oxygen reduction reaction. Appl. Catal. B. **244**, 536–545 (2019). https://doi.org/doi.org/10.1016/j.apcatb.2018.11.072
12. Jankovský, O., et al.: Synthesis of MnO, Mn_2O_3 and Mn_3O_4 nanocrystal clusters by thermal decomposition of manganese glycerolate. Ceram. Int. **41**, 595–601 (2015). https://doi.org/10.1016/j.ceramint.2014.08.108
13. Guo, F., et al.: Precursor-mediated synthesis of double-shelled V_2O_5 hollow nanospheres as cathode material for lithium-ion batteries. Cryst. Eng. Commun. **18**, 4068 (2016). https://doi.org/10.1039/c5ce01398d
14. Bartunek, V., et al.: Ultrafine ferromagnetic iron oxide nanoparticles: Facile synthesis by low temperature decomposition of iron glycerolate Perovkites-based solar cells: a review of recent progress, materials and processing methods. Mater. Chem. Phys. **180**, 272–278 (2016). https://doi.org/10.1016/j.matchemphys.2016.06.007
15. Cao, Y., et al.: Three-dimensional ultrathin In_2O_3 nanosheets with morphology-enhanced activity for amine sensing. RSC Adv. **5**, 60541–60548 (2015). https://doi.org/10.1039/c5ra09063f
16. Kwong, T,-L., Yung, K.-F.: One-step production of biodiesel through simultaneous esterification and transesterification from highly acidic unrefined feedstock over efficient and recyclable ZnO nanostar catalyst. Renew. Energy **90**, 450–457 (2016). https://doi.org/10.1016/j.renene.2016.01.028
17. Rémiás, R., et al.: Synthesis of Zinc Glycerolate Microstacks from a ZnO Nanorod Sacrificial Template. Eur. J. Inorg. Chem. **24**, 3622–3627 (2009). https://doi.org/10.1002/ejic.200900308
18. Wang, M., et al.: Layered Bimetallic Iron−Nickel Alkoxide Microspheres as High-Performance Electrocatalysts for Oxygen Evolution Reaction in Alkaline Media. ACS Sustain. Chem. Eng. **6**, 6117–6125 (2018). https://doi.org/10.1021/acssuschemeng.7b04784
19. Reyero, I., et al.: Heterogenization of the biodiesel synthesis catalysis: CaO and novel calcium compounds as transesterification catalysts. Chem. Eng. Res. Des. **92**, 1519–1530 (2014). https://doi.org/10.1016/j.cherd.2013.11.017
20. Esipovich, A., et al.: The structure, properties and transesterification catalytic activities of the calcium glyceroxide. Chem. Eng. J. **339**, 303–316 (2018). https://doi.org/10.1016/j.cej.2018.01.142
21. Sánchez-Cantú, M., et al.: Direct synthesis of calcium diglyceroxide from hydrated lime and glycerol and its evaluation in the transesterification reaction. Fuel **138**, 126–133 (2014). https://doi.org/10.1016/j.fuel.2014.08.006

22. Kouzu, M., et al.: Solid base catalysis of calcium oxide for a reaction to convert vegetable oil into biodiesel. Adv. Powder Tech. **21**, 488–494 (2010). https://doi.org/10.1016/j.apt.2010.04.007
23. Reinoso, D.M., et al.: Zinc glycerolate as a novel heterogeneous catalyst for the synthesis of fatty acid methyl esters. Appl. Catal. B. **144**, 308–316 (2014). https://doi.org/10.1016/j.apcatb.2013.07.026
24. Wan, L., et al.: Biodiesel production from soybean oil in subcritical methanol usingMnCO_3/ZnO as catalyst. Appl. Catal. B. **152**, 352–359 (2014). https://doi.org/10.1016/j.apcatb.2014.01.033
25. Wang, E., et al.: Production of biodiesel with lithium glyceroxide. Fuel **160**, 621–628 (2015). https://doi.org/10.1016/j.fuel.2015.07.101
26. Lau, P.-C., et al.: Effective heterogeneous transition metal glycerolates catalysts for one-step biodiesel production from low grade non-refined Jatropha oil and crude aqueous bioethanol. Sci. Rep. **6**, 26822 (2016). https://doi.org/10.1038/srep23822
27. Jindapon, W., et al.: Heterogeneously catalyzed transesterification of palm oil with methanol to produce biodiesel over calcined dolomite: The role of magnesium oxide. Energy Conv. Manag. **171**, 1311–1321 (2018). https://doi.org/10.1016/j.enconman.2018.06.06
28. Catarino, M., et al.: Calcium diglyceroxide as a catalyst for biodiesel production. J. Environ. Chem. Eng. **7**, 10399 (2019). https://doi.org/10.1016/j.jece.2019.103099
29. Zakharova, G.S., et al.: Preparation of TiO_2/C Composites via Titanium Glycerolate Pyrolysis. Rus J Inorg Chem. **64**, 283–289 (2019). https://doi.org/10.1134/S0036023619030227
30. Xu, J., et al.: Immobilization of zirconium-glycerolate nanowires on magnetic nanoparticles for extraction of urinary ribonucleosides. Microchim. Acta. **185**, 43 (2018). https://doi.org/10.1007/s00604-017-2596-2

Nanomaterials and Their Influence in Society Through Times

Fernando Targino Bonatto, Anne Esther Targino Bonatto, Gisele Inês Selli, and Carla Schwengber ten Caten

Abstract Nanotechnology is being identified as the source of a possible 5th industrial revolution. Also called atomic technology, it refers to a range of new technologies that operate on a nanoscale, equivalent to a billionth of a meter. This has already brought changes to society, and it will also certainly mean a substantial impact on developed and developing societies and the planet we inhabit.

Keywords Nanotechnology · Nanoscience · History · Ancient · Society

Abbreviations

B.C	Before Christ
CIT	Computer Information Technology
EDX	Energy dispersive X-ray spectroscopy
ICTs	Information and Communication Technologies
ODL	Open and distance learning
TEM	Transmission electron microscope or microscopy

1 Introduction

In 1959, physicist Richard Feynman had already started discussions about the technology for manipulating atoms and molecules, something invisible to the human eye [1, 2]. The first definition of the term nanotechnology appeared in 1974 at the Scientific University of Tokyo, proposed by the Japanese Norio Taniguchi [3], to designate

F. T. Bonatto (✉) · A. E. T. Bonatto · G. I. Selli · C. S. ten Caten
Universidade Federal Do Rio Grande Do Sul, Porto Alegre, Brazil
e-mail: bonatto02@yahoo.com.br

A. Kopp Alves (ed.), *Technological Applications of Nanomaterials*,
Engineering Materials, https://doi.org/10.1007/978-3-030-86901-4_15

a new technology beyond materials control and microscale engineering. However, the term's meaning currently comes closer to Eric Drexler's formulation, which corresponds to the processing methodology involving atom-by-atom manipulation [4].

On the other hand, nanoscience refers to the study of the phenomenon and manipulation of physical systems that produce meaningful information (i.e., noticeable differences) at a scale known as nano (10^{-9} m = 1 nm) with typical lengths that do not exceed 100 nm in length in at least one direction. In the '80s and '90s, several theories around the Japanese definition were suggested, but only from the year 2000 that this technology was effectively developed in laboratories, and soon becoming today, the center of attention of researchers, scientists, and governments around the world [2]. Therefore, the term "nanotechnology" focuses on the design, characterization, production, and application of systems and components at the nanoscale [5].

Nanotechnology has a special meaning for industrial innovation. Among so many converging sciences, nanoscience and nanotechnology appear to be an alternative for studying phenomena and material manipulation at the atomic, molecular, and macromolecular scales. On that scale, the properties differ significantly from those observed at the macro scale in terms of design, characterization, and production of structures, parts, and systems by controlling their size and shape in the nanometric scale.

Nanomaterials are present in various products of our daily lives, such as sunscreens, sports shoes, cell phones, fabrics, cosmetics, automobiles, and medicines, among others [6]. Nanotechnology is also very active and attractive in various sectors, such as energy, agriculture, water treatment and remediation, ceramics and coatings, composite materials, plastics and polymers, cosmetics, aerospace, naval and automotive, steel, dentistry, textile, cement, and concrete, microelectronics, diagnosis and prevention of diseases and drug targeting systems [7–10].

2 Ancient Nanotechnology

Considering the chemical, physical and biological processes that occur in nature, it is possible to identify the presence of nanotechnology in remote periods of human history. For example, approximately 4000 years B.C., Egyptian alchemists used the "golden elixir" to stimulate the mind and restore youth. The famous elixir of long life was actually made up of suspended gold particles with a size on the order of 1–100 nm [11].

In Europe, the color of stained glass in medieval churches, so richly worked by artisans, resulted from the formulation of glass with gold nanoparticles [12, 13]. The decoration of churches with colored stained glass was common—in addition to creating an atmosphere suitable for contemplation, it was considered a form of catechesis. The addition of gold chloride to the molten glass gave rise to gold spheres of approximately 25 nm in diameter, which, due to the scattering of light, generated a red color. In the nineteenth century, Michael Faraday showed the relationship between the properties and size of gold particles, noting that this size influenced the absorption of light [13, 14]. In this way, it is possible to obtain gold-based materials in different colors, depending on the size of the particles. In its natural bulk form, gold exhibits a yellow coloration. However, depending on the size of the particles, they can appear black, ruby, or purplish. Other colors could be obtained by changing the material used and/or the particle size. For 50 nm gold nanoparticles, for example, a green tint appears. If the size is doubled (reach 100 nm), the scattered color becomes orange, while 100 nm silver nanoparticles scatter yellow light. Blue color can be obtained with 40 nm silver nanoparticles [14, 15] (Fig. 1).

Another notable example is the Indian ink, developed by the Chinese around 2700 B.C., and consists of carbon nanoparticles suspended in an aqueous solution [16]. Although typically nanoparticles dissolved in a liquid aggregate, forming micro and

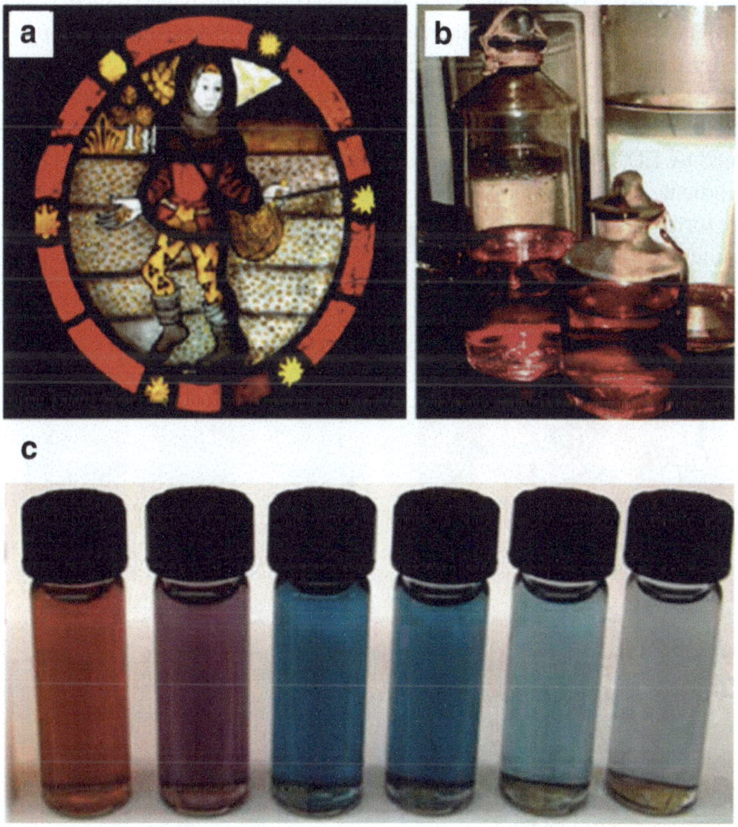

Fig. 1 Examples of Au in several different formats: **a** nanoparticles in ancient church windows as indicated by the striking red stain, **b** nanoparticles in Faraday's colloidal samples, and **c** nanocages with different wall thicknesses and thus distinctive colors. Reprinted with permission from Yang et al. [13]. Copyright 2015 American Chemical Society

macroparticles that tend to precipitate, separating from the liquid, the ancient Chinese discovered that it was possible to stabilize India ink by mixing a glue (acacia gum) into the solution with coal powder and water. Today it is possible to understand that, by binding to the surface of the carbon nanoparticles, the glue molecules prevent their aggregation and, therefore, their separation from the liquid core [17].

Perhaps the most cited example of ancient nanotechnology is the Lycurgus Cup (Fig. 2) [18]. With 15 cm high and 700 g of glass and silver, the cup has been owned by the Rothschild family since the mid-nineteenth century, being acquired by the British museum in 1958. This is the only complete example of Roman glass that presents dichroism; that is, the material has two colors: green when reflecting and red when transmitting light. From the 1950s onwards, the Lycurgus Cup began to attract the attention of the scientific community.

In the 1960s, a hypothesis was raised that metallic colloidal dispersions could explain the effect produced by the cup in the glass. However, this hypothesis was only confirmed in 1990 by Barber and Freestone [19]. They published a study in which they analyzed glass fragments of the chalice by transmission electron microscopy (TEM) and energy-dispersive X-ray spectroscopy (EDX). By TEM, the authors were able to identify nanometric materials dispersed in the glass matrix with lengths between 50 and 100 nm. The characterization of the elements present in the particles was performed by EDX, a technique derived from X-ray fluorescence spectroscopy. The results indicated that the nanoparticles were mainly gold and silver, both in the alloy and pure form. In addition to these, smaller particles (up to 15 nm) of sodium chloride were found.

Fig. 2 Lycurgus Cup: the first example of man-made glass nanocomposite. Reproduced with permission from Nur et al. [18]

Freestone et al. [20] report evidence that, despite being rare, other objects like this could have existed, according to letters from the third century. However, although the exact process of production of the chalice is unknown, the authors present some assumptions that range from accidents with gold leaf in the molten glass to the reuse of glass pigmented with gold and silver. The authors also state that it is accepted that most probably the Roman artisans did not know that it was the proportion of gold and silver dispersed in the glass that caused this effect, considering that the technology had a very short life, not surpassing the fourth century.

3 Nanotechnology in Modern Society

Nanotechnology enabled the emergence of new products marketed by companies founded in the new millennium, obtaining millionaire annual revenues. In this sense, several international companies operate in the nanotechnology sector in various fields, especially in nanoelectronics and in the cosmetics area. Nanoelectronics revolutionized industrial parks and, in synergy, with the advancement of communication and information technologies (CIT), there was the possibility of the emergence of new markets, new professions, new products, and services. The globalization process, advances in computer science, and nanotechnological transformations have made the market more fluid and constantly changing. The World Economic Forum, held in Davos in January 2018, discussed one research that estimated that 65% of students starting elementary school today will work in professions that do not yet exist.

From a pedagogical point of view, active methodologies and communication and information technologies (ICTs) in the teaching–learning process are increasingly frequent. In this sense, nanoelectronics in synergy with programming and computer science are affecting education from the most basic use of technology, in an expository way with the help of multimedia projectors and interactive whiteboards in the classroom mobile technologies and interactive in both basic and higher education.

In addition, nanoscience has allowed the advancement of technologies such as virtual reality, simulators, and even technologies already standard in classrooms, such as the projection of slides, videos, images, and texts, which are extremely important to streamline and improve the process of teaching–learning in all academic areas. The use of active methodologies in games and group dynamics use applications on mobile devices and even open and distance learning (ODL) or hybrid learning.

One of the main areas of impact in the field of medicine and health occurs in pharmaceutical drugs. This is because the nanoengineered medicines can be transported and released in the body according to the need for the diseased tissue or organ. This applies from analgesics, anti-inflammatory drugs, and cosmetics to antitumor drugs in the research phase already present in the pharmaceutical market.

In addition to drugs, protheses with nanocoatings and made of nanomaterials such as nanocomposites can increase biocompatibility and change the physical–chemical properties of materials, causing the patient's immune system to recognize that material external to the body as something from the body itself and reduce the chances of rejections after surgery [21, 22].

In the context of food engineering and agriculture linked to bionanotechnology, packages composed of nanofilms that preserve food for a longer time and change its color through decay or pathogenic microbiological contamination are already realities in the literature [23–25].

Finally, it is noteworthy that the disposal of nanostructured materials with their growing use in industrial processes and products, including the oil and gas industry, promotes the growth of 'nano-waste', which has consequences that are not yet very clear to ecosystems [26–29].

The impact of nanotechnology on society occurs nowadays, mainly in the context of nanoelectronics concerning mobile technologies (smartphones, notebooks, tablets, among others) that caused several socio-cultural transformations in civilization in its last decades, and this includes the rapid changes in generations after globalization with scientific-technological advances and its easy access.

In addition, it is noteworthy that the various nanotechnologies that make Industry 4.0 will generate changes in ethical and behavioral values, in addition to causing a high impact on professions by extinguishing certain functions and training while providing for the emergence of others [30]. Understanding the implications of significant social transformations caused by technologies, experts such as Schwab [31, 32] point out that the convergence of technologies, the creation of physical and virtual data flow control networks, and the provision of interconnected global services bring some political risks. Social relations regarding the speed at which these contexts transform and reshape social relations in dissonance with the social subject itself, the breadth and depth that these systems can bring with the changes in the production of consumer goods, and the systemic impact that a revolution of exponential magnitude can cause in the world [33].

References

1. Editorial: 'Plenty of room' revisited. Nat. Nanotechnol. **4**(12). 781 (2009). Nature Publishing Group. https://doi.org/10.1038/nnano.2009.356
2. Bayda, S., Adeel, M., Tuccinardi, T., Cordani, M., Rizzolio, F.: The history of nanoscience and nanotechnology: from chemical-physical applications to nanomedicine. Molecules **25**(1), 112 (2020). https://doi.org/10.3390/MOLECULES25010112
3. Wennersten, R., Fidler, J., Spitsyna, A.: Nanotechnology: A new technological revolution in the 21st century. Handb. Performability Eng. 943–952 (2008). https://doi.org/10.1007/978-1-84800-131-2_57
4. Drexler, K.E.: Molecular engineering: An approach to the development of general capabilities for molecular manipulation. Proc. Natl. Acad. Sci. U. S. A. **78**(9), 5275 (1981). https://doi.org/10.1073/PNAS.78.9.5275
5. Sharma, P.K. et al.: Nanotechnology and its application: a review. Nanotechnol. Cancer Manag. 1–33 (2021). https://doi.org/10.1016/B978-0-12-818154-6.00010-X
6. Ramsden, J.J.: Applied nanotechnology: The conversion of research results to products: second edition. Appl. Nanotechnol. Convers. Res. Results to Prod. Second Ed. 1–217 (2013). https://doi.org/10.1016/C2012-0-06410-1

7. Encyclopedia of Nanotechnology. Encycl. Nanotechnol. (2012). https://doi.org/10.1007/978-90-481-9751-4
8. Prasad, R., Bhattacharyya, A., Nguyen, Q.D.: Nanotechnology in sustainable agriculture: recent developments, challenges, and perspectives. Front. Microbiol. 8(June) (2017). https://doi.org/10.3389/FMICB.2017.01014
9. Raj, S., Jose, S., Sumod, U.S., Sabitha, M.: Nanotechnology in cosmetics: Opportunities and challenges. J. Pharm. Bioallied Sci. **4**(3), 186 (2012). https://doi.org/10.4103/0975-7406.99016
10. Papadaki, D., Kiriakidis, G., Soutsos, T.T: Applications of nanotechnology in construction industry. Fundam. Nanoparticles Classif. Synth. Methods, Prop. Charact., 343–370 (2018). https://doi.org/10.1016/B978-0-323-51255-8.00011-2
11. Charlier, P., et al.: A gold elixir of youth in the 16th century French court. BMJ **339**(7735), 1402 (2009). https://doi.org/10.1136/BMJ.B5311
12. Louis, C.: Gold Nanoparticles in the Past: Before the Nanotechnology Era (2017). Accessed 20 July 2021. [Online]. Available: www.worldscientific.com
13. Yang, X., Yang, M., Pang, B., Vara, M., Xia, Y.: Gold nanomaterials at work in biomedicine. Chem. Rev. **115**(19), 10410–10488 (2015). https://doi.org/10.1021/ACS.CHEMREV.5B00193
14. Pluchery, O., Remita, H., Schaming, D.: Demonstrative experiments about gold nanoparticles and nanofilms: an introduction to nanoscience. Gold Bull. 464, **46**(4), 319–327 (2013). https://doi.org/10.1007/S13404-013-0122-9
15. Pfeiffer, C., Parak, W.J., Montenegro, J.M.: Synthesis of colloidal gold and silver nanoparticles and their properties. Bio- Bioinspired Nanomater. **9783527335817**, 1–22 (2014). https://doi.org/10.1002/9783527675821.CH01
16. Binns, C.: Introduction to Nanoscience and Nanotechnology, p. 301. Wiley, New York (2010)
17. Sanchez, C., et al.: Acacia gum: History of the future. Food Hydrocoll. **78**, 140–160 (2017). https://doi.org/10.1016/j.foodhyd.2017.04.008ï
18. Nur, O., Willander, M.: Low temperature chemical nanofabrication, low temp. Chem. Nanofabrication (2020). https://doi.org/10.1016/C2016-0-03914-1
19. Barber, D.J., Freestone, I.C.: An investigation of the origin of the colour of the Lycurgus cup by analytical transmission electron microscopy. Archaeometry **32**(1), 33–45 (1990). https://doi.org/10.1111/j.1475-4754.1990.tb01079.x
20. Freestone, I., Meeks, N., Sax, M., Higgitt, C.: The Lycurgus Cup—a roman nanotechnology. Gold Bull. 2007 404 **40**(4), 270–277 (2007). https://doi.org/10.1007/BF03215599
21. Gallo, J. et al.: Silver nanocoating technology in the prevention of prosthetic joint infection. Materials **9**(5) (2016). https://doi.org/10.3390/ma9050337
22. Lim, P.N., Chang, L., Thian, E.S.: Development of nanosized silver-substituted apatite for biomedical applications: A review, nanomedicine: nanotechnology. Biol. Med. **11**(6), 1331–1344 (2015). https://doi.org/10.1016/j.nano.2015.03.016
23. Rolim H.M.L., Ramalho, T.C.: Biopolymer essential oil nanocomposite for antimicrobial packaging. In: Biopolymer-Based Nano Films, pp. 115–131 (2021)
24. Rai, M., Yadav, A., Ingle, A., dos Santos, C.A.: An introduction to biopolymer-based nanofilms, their applications, and limitations. In: Biopolymer-Based Nano Films, pp. 3–17 (2021)
25. Zambrano-Zaragoza, M.L., González-Reza, R.M., Quintanar-Guerrero, D., Mendoza-Muñoz, N.: Nano-films for food packaging. Food Eng. Ser. 287–307 (2020). https://doi.org/10.1007/978-3-030-44552-2_10
26. Musee, N.: Nanowastes and the environment: Potential new waste management paradigm. Environ. Int. **37**(1), 112–128 (2011). https://doi.org/10.1016/j.envint.2010.08.005
27. Part, F., Zecha, G., Causon, T., Sinner, E.K., Huber-Humer, M.: Current limitations and challenges in nanowaste detection, characterisation and monitoring. Waste Manag. **43**, 407–420 (2015). https://doi.org/10.1016/j.wasman.2015.05.035
28. Bouillard, J.X. et al.: Nanosafety by design: Risks from nanocomposite/nanowaste combustion. J. Nanoparticle Res. **15**(4) (2013). https://doi.org/10.1007/S11051-013-1519-3
29. Campos, A., López, I.: Current status and perspectives in nanowaste management. Handb. Environ. Mater. Manag. 2287–2314 (2019). https://doi.org/10.1007/978-3-319-73645-7_161

30. Xu, M., David, J., S. K.-I.: Journal of financial, and undefined 2018. The fourth industrial revolution: Opportunities and challenges. *creo.sc-celje.si*, Accessed 27 July 2021. [Online]. Available: http://creo.sc-celje.si/pluginfile.php/3889/mod_resource/content/1/4.1.4_01_The fourth industrial revolution.pdf.
31. Lund, B.: The fourth industrial revolution. Inf. Technol. Lib. 40(1) (2021). https://doi.org/10.6017/ITAL.V40I1.13193
32. Klaus, S.: The fourth industrial revolution|foreign affairs. Foreign Affairs (2015). https://www.foreignaffairs.com/articles/2015-12-12/fourth-industrial-revolution . Accessed 27 July 2021
33. Maynard, A.D.: Navigating the fourth industrial revolution. Nat. Nanotechnol. **10**(12), 1005–1006 (2015). https://doi.org/10.1038/nnano.2015.286

The manufacturer's authorised representative in the EU is Springer Nature Customer Service Centre GmbH, Europaplatz 3, 69115 Heidelberg, Germany. If you have any concerns regarding our products, please contact ProductSafety@springernature.com

Printed and bound by CPI Group (UK) Ltd, Croydon, CR0 4YY
30/09/2026
02225695-0003